PSYCHEDELIC AMERICAN

VOL. 1, ISSUE 1

DISCLAIMER

This journal is intended as an informational guide for educational and philosophical purposes. Neither the authors nor the publisher encourage, endorse, or support illegal or dangerous behavior of any kind. Readers assume full responsibility for their choices and actions, including but not limited to, any physical, psychological, or social consequences resulting from the ingestion of psychedelic substances or their derivatives.

Table Of Contents

Editor's Note
Benjamin Stolz
Editorial Director for Psychedelic American

Greetings, and thank you for your interest in Psychedelic American! After several months of hard work, our premier issue is finally ready to hit the shelves, and we are ready for public readership to begin.

Late in the summer of 2014, I dreamed up a publication that would bridge the gap between the generations of psychedelic artists, scientists, critics, journalists, researchers, and thinkers alike. With lots of community support, we were finally able to make it happen. The final platform is an open journal which intends to be entertaining as well as thought provoking, and which hopes to shed light on the contributors' unique approaches to their subject matter more than anything else.

We are at a crossroads in the history of psychedelic drugs and culture in the United States. The interactions, sometimes better described clashes, between major forms of media in the 21st century has created an environment which is very fertile for discussion, similar to that which spawned the counterculture of the 1960s. As mainstream television struggles between its traditional propaganda memes and the more recent development of the "psychedelic renaissance," the internet communities are busy searching for the next evolutionary steps in creating a unique, authentic psychedelic pulsar for the Digital Age. My hope is that Psychedelic American will play an integral role in shaping the themes of the new media and creating dialogue that reaches out across generations and demographics.

The publication will be available in both digital (with photos) and print (no photos) formats. Four issues are planned per year, and subscriptions will be available starting with the second issue. A forum, which is already being hosted on Facebook, and a newsletter to compliment the first issue are also in the works.

Lastly, I would like to extend my deepest appreciation and gratitude to everyone who has contributed thus far to Psychedelic American, but especially I would like to thank Tristan Gulliford, Randy Sloane, and Angel Majao for their dedication to the project, as well as James W. Jesso and Eduardo Mainero for their generosity in organizing the publication of this issue.

Regards,
Benjamin Stolz

Whatever Happened to 2012? Part 1

Preview by Tristan Gulliford

I first heard about 2012 around 2002, about a decade before the prophetic events we were told would occur. Through psychedelic author Daniel Pinchbeck's *Breaking Open the Head* online forum, I became an early adopter of the 2012 prophecy meme. I spent years daydreaming about the glorious metaphysical supernatural happenings which were close upon us in those days. I spent hours talking with people about 2012 and what they believed might happen. I got caught up in fervent spiritual hype around exotic ideas based on esoteric interpretations of foreign cultures' mystical writings and art. My belief and hopes for the future cast a shadow on my critical rationality, leading me to confuse my inner subjective reality with the objective outer reality.

However, by the time 2012 actually came around, I was already burned out on the idea and my belief had essentially dried up, because I had not seen any signs that this prophecy would come to pass. So I was not surprised when nothing happened, but what did surprise me was when some of my friends pretended as if "something" had happened, and there was no reason to question their previous beliefs, when the solstice of December 21st, 2012 came and passed without so much as a whimper.

No earthquakes, no apocalypse, no death. No utopian merging of cultures into an ultimate rainbow tribe. No super-orgasmic collective worldwide DMT trip. No Ending of Time into an ingressed spiral of concresced novelty. Just a normal Friday, which became Saturday, Sunday, and then we all went back to our everyday lives on Monday. We all partied hard for the weekend, with lots of spiritually enlightened costumes, body paint, crystals, and bass music but no one really knew why. So, whatever happened to 2012?

To say that the visions we were promised for 2012 were grandiose is a supreme understatement. We were promised nothing short of a world-changing event that would affect not only us, but the laws of time and physics, the natural laws of science. Speculations as to what would happen upon that crucial transformation date of Dec.21st, 2012 were as diverse and imaginative as the crowd who conceived them:

- The Mayan calendar predicted a transformation of consciousness, and we would witness a species-wide uplifting that would create a spiritual utopian society based on the values of compassion and harmony.
- Aliens would descend from the skies and welcome us into an intergalactic brotherhood of peace.
- Solar flares would cause apocalyptic natural disasters which would destroy society and force us to rebuild according to the guidelines of indigenous wisdom traditions.
- The Earth's electromagnetic or "etheric" poles would shift, resulting in disruptive technological imbalances and changes in human consciousness.
- Galactic alignment with a central black hole would cause a dimensional shift to occur, after which people would be more directly affected by things such as cosmic wave radiation, altering their consciousness.
- Planet X would smash into earth's orbit, sending us off-kilter and causing multiple cascading natural disasters.
- A new spiritually-based time calendar would be adopted which would solve most of the world's modern problems.

- Time would end, novelty would reach an infinite end-point, and the laws of physics change so drastically that human life would no longer resemble anything that had come before.

None of these things actually happened on Dec. 21st, 2012, or any time since then. So, what really did happen? How did hundreds of thousands of people come to associate with a prophetic belief that appears to have been grounded in nothing substantial at all? What a fascinating moment to analyze human psychology: how and why we are motivated to join groups with similar belief structures, and how collective beliefs can shape the individual mind. However, most people have unfortunately missed this important opportunity for self-reflection, and merely continued along the same path without asking questions about how they got there.

The origins of the 2012 prophecy meme begin with Yale archaeologist Michael Coe's book *The Maya* (1966), who, eerily enough, was a Korean War-era spy. Southwest author Frank Waters wrote about prophetic ideas related to the Mayan calendar in his book *Mexico Mystique: The Coming Sixth World of Consciousness* (1975). The first mention of the prophecy meme in popular culture was on the Leonard Nimoy-hosted "In Search Of... The Mayans" in 1978.

> "The ancient Mayans, men of knowledge, conceived their time on earth, their cycle of civilization, to be 5,200 years, beginning their calendar August 12th, 3113 B.C. They predicted that on December 24th, 2011 A.D. a cataclysmic earthquake would terminate their cycle of civilization. New men of knowledge would then appear to fight the forces of evil, and lead the people to create a world government. If the Mayan men of knowledge were right, in just thirty-four years we may learn the answers to some of the ancient Mayan mysteries."

-Leonard Nimoy, "In Search Of... The Mayans", 1978.

Jose Arguelles was the first New Age author to popularize the 2012 prophecy. His book *The Mayan Factor: Path Beyond Technology* written in 1987 set the stage for what would develop, over the course of more than 20 years, into a mainstream cultural phenomenon with associated Hollywood blockbuster movies and counterculture icons.

Psychedelic author and lecturer Terence McKenna further elaborated the apparently complicated eschatology of 2012 prophecy in books like *The Invisible Landscape* and *The Archaic Revival,* and in countless lectures.

> "What is happening to our world is ingression of novelty toward what Whitehead called 'concrescence,' a tightening gyre. Everything is flowing together. The 'autopoetic lapis,' the alchemical stone at the end of time, coalesces when everything flows together. When the laws of physics are obviated, the universe disappears, and what is left is the tightly bound plenum, the monad, able to express itself for itself, rather than only able to cast a shadow into *physis* as its reflection. I come very close here to classical millenarian and apocalyptic thought in my view of the rate at which change is accelerating. From the way the gyre is tightening, I predict that concrescence will occur soon—around 2012 A.D. It will be the entry of our species into hyperspace, but it will appear to be the end of physical laws accompanied by the release of the mind into the imagination."
>
> Terence McKenna, *The Archaic Revival* (p. 101).

Needless to say, physical laws are exactly the same now as they were before 2012, and have likely been for billions of years prior. No such concrescence of anything actually culminated at this time. And as much as we might like to pretend that we do, we don't live inside our subjective imaginations. Objective reality is still real, and the rules that bind us here haven't changed.

Author Daniel Pinchbeck picked up the torch where McKenna left off, writing the book *2012: The Return of Quetzalcoatl* (2006) in which he claimed to be the avatar of the Aztec god Quetzalcoatl, as well as a reincarnation of Indian emperor Ashoka, who ruled the Mauryan Empire from 272 to 232 B.C. (p. 370-372). Here are some excerpts from the prophetic passage near the end of the 2012 book in which Pinchbeck reveals the revelatory prophecy he "received" while drinking ayahuasca:

> "I am an avatar and messenger sent at the end of a kalpa, a world age, to bring a new dispensation for humanity—a new covenant, and a new consciousness [...] Soon there will be a great change to your world. The material reality that surrounds you is beginning to crack apart, and with it all of your illusions. The global capitalist system that is currently devouring your planetary resources will soon self-destruct, leaving many of you bereft [...] You are, right now, living at the time of revelation, Apocalypse, and the fulfillment of prophecy. Let there be no doubt [...] What you are currently experiencing as the accelerated evolution of technology can now be recognized for what it is: a transition between two forms of consciousness, and two planetary states [...] The End of Time approaches. The return of Quetzalcoatl foreshadows the imminent closing of the cycle and the completion of the Great Work".
> Daniel Pinchbeck, *2012: The Return of Quetzalcoatl* (p. 367-370).

What were the effects of all these grandiose end-of-the-world, beginning-of-a-new-one messages? For some people there were serious consequences. I talked to a girl who believed that she should get two or three credit cards and max them out, move to Bali, and wait out the end of the world, since the debt wouldn't matter after 2012. I can only hope that people did not follow through on these kind of desperate alarmist plans, but I'm sure that some did. Media noted that at least one suicide occurred on December 21st related to apocalyptic beliefs. An Ipsos poll showed that 10% of people polled worldwide believed that 2012 marked an apocalyptic event.

There are many levels of irony in the 2012 prophecy: one of the most obvious is comparison to evangelical end-times prophecy. Evangelical preacher Harold Camping predicted an apocalypse for May 21st, 2011 which failed to materialize. Many of the liberal believers in 2012 likely mocked Christian prophets like Camping, while simultaneously holding fast to their own apocalyptic 2012 beliefs, which came from different sources. This cognitive dissonance was not discussed in the New Age culture, despite the prophecy meme having most of the same millennialist trappings as evangelical Christian prophecies. Well-educated people who generally consider the evangelical prophecies puerile and delusional, nonetheless held onto beliefs about the transformation of 2012 for years. How does this happen?

Even within the apparently renegade underground of the "counterculture" sub-culture, beliefs, opinions, tastes, attitudes, styles of dress, linguistic patterns, etc. are all shared, repeated, and reinforced to form a coherent and recognizable aesthetic. This is how conformity creates participative group identity. The more we wanted 2012 to happen, the harder we believed something would happen, and the more we mimetically spread that belief to others and reinforced it within our own immediate social groups. This went on for years.

What has disappointed me most in the aftermath of failed prophecy was not that the prophecy itself failed, but that nobody seemed upset about it. We carried on as if nothing had happened, as if we'd never expected anything to change. The cohesiveness and tangible social value we get from inclusion in the group has apparently overruled the rational mind. If this is the case, are we no better than the Harold Camping's and Michele Bachmann's of the world? Are we buying into a different flavor of religious fundamentalism? Have we developed our own psychedelically-induced version of the Christian messiah complex?

One grievous error made in the search for 2012 prophecy, which the New Age should own up to and recognize as a critical mistake that deserves some atonement, is the cultural misappropriation of the Mayan calendar and other indigenous wisdom teachings. Despite Terence McKenna's clear articulation of the "Other" and "Self", we seem to have externalized the Other, at least since McKenna's death in 2000, onto the Mayan people, giving them magical primitive powers beyond our comprehension as modern Americans but nonetheless boldly interpreting their sacred writings into our own cultural context which is vastly removed from the cultural context of the classical Maya.

This attitude is now called "Mayanism," as the late 20th century paternalistic fetishization and exoticized projection onto Asian culture is now called "Orientalism". While some brief apologies were made for this after the prophecy failed, nothing came close to addressing the serious missteps in judgment that were required for such rampant "Mayanism" and ultimately the disrespect and misinterpretation of another culture's sacred mythology. I, for one, hope we learn from this and do not make the same mistakes again.

In order to put protective measures into place, we should first recognize that different world cultures are not our own, and it may not be so easy or simple to grab parts of that culture that we want and integrate them into our modern context while not understanding a larger subjective context of belief that can only come from being raised inside that culture. To assume we can have insider knowledge of a foreign culture is offensive, and almost always wrong.

Critical thinking is crucial. When someone wants to sell you an idea you should demand proof that this idea is worth investing in. Tangible, quantifiable proof. If the only proof you can find is a "feeling," well then the idea is probably bullshit. Your intuition and rationality need to work hand in hand to parse out truth from fiction. Demand results from the ideas and beliefs that you choose to attach yourself to and spread to others. Question authority and do not assume that someone knows what they're talking about just because they sound good when they speak. Go beyond rhetoric and beyond emotion, look at the facts to find truth.

Finally, we should develop an appropriate context for celebration, and an appropriate context for work. If we want to realize the hopes and dreams that we all projected into 2012, how will we best achieve this? Creating an intimate and insular counterculture community helps us feel special and privileged, but to actually change the whole world like we wanted to with 2012 will require far more focus and effort directed outwards, towards conscious activism: not just aligning our intentions towards a positive end goal, but aligning our intentions and then following through with proper action to achieve that positive goal.

Whatever Happened to 2012? Part 2:

Psychedelic American (PsyAm) in conversation with Tristan Gulliford (TG), John Hoopes, (JH) and Kevin Whitesides (KW).

PsyAm: JH, your path to studying this meme was a bit more indirect than our other two interviewees...

JH: Even though my undergraduate advisor was Michael Coe (at Yale), I didn't pay much attention to 2012 until I went to Burning Man in 2002 and saw Daniel Pinchbeck. By next year I was invited to add a few words to Lorenzo Haggerty and Daniel Pinchbeck's presentation. Already a long-time internet and PC user, I began participating in the Breaking Open the Head forums, where I met Tristan (under a username). The 2012 meme became something of interest when it started "plugging in" to other issues in the counterculture...

PsyAm: What are the biggest misconceptions about the Mayan Calendar?

JH: The single biggest one is that the Maya predicted the end of reality, or the world, or the calendar.   There are no Mayan doomsday prophecies and never were. Looking at the key date of the turn of the 13th b'ak'tun (lunar cycle), there is no evidence that the Maya believed this b'ak'tun to be more significant than any other.

PsyAm: Most of this comes from the Goodman (properly called the Goodman-Martinez) set of correlations...can you talk a bit about that?

JH: Goodman had two ideas that have not held up to scrutiny. One was a 13-b'ak'tun cycle. The other was a unique galactic alignment, which astronomers have found no real evidence to support either....

PsyAm: Does the New Age constantly require a re-invention of an eschatological mythos?

JH: Remember that this is where the term "New Age" comes from in the first place. The major transition takes place on a Broadway stage in the late 1960s, when the musical Hair opens with the chorus of "This is the dawning of the Age of Aquarius." Very soon after, the mythology about 2012 replaced the New Age. There is some debate as to whether or not one is the other, but I see 2012 as having moved in and occupied the same place in the discourse that the Age of Aquarius did in the 1960s and early 1970s.

KW: I have been arguing recently that in order for something to be New Age, it has to revolve mainly around millennialism, and eschatology. Michael Barkun has a term for what happened to the 2012 meme, called "improvisational millennialism" where the participant picks and chooses what best suits their narrative. No two tellings of 2012 wind up the same, and when one author borrows from another, they are further involved in the same process of borrowing. When Michael Coe writes in 1966 about Armageddon and the Long Count, he was doing it as a scholar, who had lots of comments and observations. By the time Frank Waters got his hands on the same bit of information, he transformed the story into one involving Atlantis and UFOs...none of which was in Michael Coe's narrative. It wasn't any sort of ideological motive, it was just a buffet-style grab.

PsyAm: As 2012 approached, what happened?

KW: As far back as the 1970s, there were popular documentaries narrated by Rod Serling and Leonard Nimoy about 2012, about, you know, ancient aliens and Armageddon. And this telling of the meme in popular culture stuck into the X

Files, whose last episode "revealed" that the Mayans were the old shadow government in 2002...

PsyAm: So this was what the mainstream was experiencing...

KW: I ask myself, "to what degree did this all stick?" Do people retain this meme, who pick it up in popular media? Where I think it began to stick was in 2006, when we started getting History Channel documentaries like "Doomsday 2012." I actually did an analysis of around a dozen of the 2012 television documentaries, and found that 80-90% carried this first "doomsday" narrative, while only around 10% bore the "transformation of consciousness" narrative that the counterculture was experiencing. So really there were two very different messages being sent (as 2012 approached.)

PsyAm: There were a lot of misconceptions about how Terence McKenna arrived at his conclusions about 2012, using his Timewave Zero program...

KW: There's a foreword that he wrote on a Phillip K Dick collection somewhere...where he notes that the first date he picked for the eschaton was his next birthday, because he felt he had to be of central importance. He writes about sitting on a cushion on his birthday waiting for the world to unravel around him. He changed the end date around a few more times, one having to do with the death of his mother...

TG: It's mathematically arbitrary!

KW: [Rupert] Sheldrake and [Ralph] Abraham, in one of the Trialogues talks, agree that it's an occult dogma – it requires Terence to interpret it. He really had no way of scaling his wave over time, but knew there was an eschatological moment implicit in there.

PsyAm: I see. Interesting way of putting it...

KW: He says in one recording that the eschaton would almost certainly come in "our lifetime" and it's no surprise he settled on the Hiroshima bombing as a significant date, which put the eschaton into November 2012. McKenna didn't remember who first told him about the Long Count Calendar, it may have been Henry Munn, but at the 1985 Ojai conference he does assert that he changed the date to the winter solstice of 2012 after he found out about the Maya.

PsyAm: How did you get involved in 2012, TG?

TG: I got involved around 2002, when **JH** and I joined the Breaking Open the Head forums. It was really in vogue when [Daniel] Pinchbeck's book came out, and I remember standing-room only tents at Burning Man packed to hear him speak in 2006. I got very cynical about all this around 2010, so I didn't experience 2012 the way most of my peers did. By that time, I was looking around looking for tangible changes, and explaining to my friends that all of that energy, focus, and faith would have been better spent elsewhere...

PsyAm: You were studying religion at the University of Colorado at the time you were involved. Do you feel there was a cult aspect of 2012 that the counterculture is already coming to regret?

TG: I had never thought about it when the stuff was going on, but it seems to me that the counterculture has eschatological beliefs that are no different from the modern evangelical prophecies, which are proven wrong time and time again.

PsyAm: No one's walked out of either church...

TG: My friends and I were intelligent, liberal people who would make fun of these evangelicals, who we laugh at because they're not rational thinkers, and yet we were going along with this thing that never happened, like the "Emperor's New Clothes." I would say it had the characteristics of a cargo cult; as the original picture became farther away in the distance, the interpretations went more and more all over the place...

KW: Following that lead, I would note how we're dealing with something which went from a paragraph in Coe's book in 1966 about the Long Count Calendar to a rap about Armageddon in the works that followed...

TG: If you add up how many people believed in some narrative of 2012, it is a shockingly large number.

PsyAm: It covered a spectrum ranging from Burning Man to the Alex Jones radio show, the latter of which led to some 2012-related violence...

KW: Do you know who Ramtha is? She's a channeler with a big school in Washington, who had a few students from her school wanted to build a bunker, and ended up in a shootout with police...

TG: (Laughs) Sounds like Jonestown!

PsyAm: Yeah, I guess the message here is that violence happens on the left wing just as easily.

TG: I talked to a girl who was so convinced of her own 2012 narrative that she was prepared to max out her credit cards and move to Bali to wait for an apocalypse, and I have to wonder how many people there are out there that did, or almost did, follow through on those sorts of things when the time came...

PsyAm: Well, there was one telling of the narrative that *did* emphasize moving to remote or exotic locations, on the right wing...that a virus released by the New World Order would kill us all, etc. etc. ... once we get past the pseudoarchaeological doomsday memes and psychedelic-influenced memes, we get into the really bizarre improvisations...once again, which seem to cover the entire cultural-countercultural spectrum.

TG: My question to Kevin is what parallel can be drawn between 2012 and Christianity, specifically the prophecies of the evangelicals...

KW: There are a lot of varieties in the ways that people thought about what would happen in 2012. A lot of it did reflect the notion of the "end times" and prophetic notions, but I would say the key to your question would be to gauge each individual, you know, how evangelical *is* this person?

TG: Oh [my group of friends and I] were certainly evangelical, especially when mushrooms were lying around...

PsyAm: What is the common nature of the 2012 meme? Is it even best characterized as a meme?

JH: I like Johann Normark's interpretation [of 2012]...he calls it a hyper-object, with its own body of mythology, in which a variety of different things are reflected. The mythology can be interpreted in many different ways by many different people, who are connected by this feeling that we are in the midst of a metaphysical transformation of sorts. Soon the whole thing took on a life of its own...I've also referred to 2012 as a "secret handshake."

PsyAm: What do you think is the source of the improvisational nature of the 2012 phenomenon?

JH: Virtually every telling of 2012 had some element of embellishment to it, but virtually all drew upon ideas that had been batted around in spirituality communities in the West for years. I point to Madam Blavatsky and the Theosophists as being particularly important in pulling East and West together. The Maya are there as this motif of mystery (laughs) and I guess just a way to color the whole thing with a mysterious cast, beyond most people's understanding.

PsyAm: What role did Burning Man play in spreading the 2012 meme?

TG: I was there for a lot of Pinchbeck's talks. The word "faith" comes to mind again, because he was being treated...like a prophet. We basically sat there and waited for him to interpret the future for us. We were going to venture forward into this telepathic, utopian society beyond materialism....

PsyAm: Did 2012 need a living prophet, as the date approached?

TG: I mean, it was Daniel who brought it into the pages of Rolling Stone...did he have to claim that he was a misappropriated Aztec god, talking about a supposedly Mayan prophecy? One of the things that Terence McKenna said that I still agree with was that we shouldn't exoticize the Other, because there is no difference between the Other and ourselves...but that's exactly what happened.

PsyAm: As we know, nothing happened on December 21, 2012. The laws of physics didn't change, no mystery planets collided, and no endless DMT Trip at the end of time...

KW: (Laughs) If you began to think that the psychedelic experience was connecting you to these higher realities, that you wanted everyone to experience, why wouldn't the [drugs] begin to play a larger role in the narrative?

TG: The meme for a lot of people did become as stupid as, "wouldn't it be cool if everyone tripped at the same time, in perfect utopian fashion?" And it didn't even seem ridiculous at the time.

PsyAm: What happened after December 21, 2012, and where do we go from here?

TG: It was another "Emperor's New Clothes" situation. Everyone has apparently gotten so invested in their place in the counterculture that they aren't willing to take responsibility for such a major failure. If we are going to demand an apology, it shouldn't necessarily be for all the reckless behavior, but rather for completely disrespecting another culture when we all knew better.

KW: As a scholar of 2012, coming from an academic perspective, I don't see it going anywhere. There were hundreds of books published in the decade or so leading up to it, and films, and TV shows, and none of the popularity seemed to falter until the actual date of December 21 came and went...it became so popular that I don't think there will enough time to wrap my head around all of it!

MDMA, PTSD And The Future Of Psychiatry

Dr. Ben Sessa
MBBS BSc MRCPsych
Consultant Psychiatrist in Substance Misuse and
Psychedelic Researcher, Cardiff University Medical School, UK

Introduction

Psychiatry is in desperate need of reformation. The stranglehold over medical research by the pharmaceutical industry is a consumer-driven distraction that insists on daily dosing with drugs aimed only at masking symptoms. But what patients with trauma-based mental disorders really need is guided, focused, intensive psychotherapy that allows them to safely re-visit their painful memories. However this form of therapy is unpopular with an industry that expects a sellable 'product' at the end of expensive Research and Development programs. Whilst the pharmaceutical industry continues to rake in the profits of millions of patients on long-term drug therapies that do not bring about lasting remission, who is going to pay for the hugely expensive research studies that need to be done with psychedelic drugs? Are we ever going to see these medicines licensed as mainstream psychiatric drugs? (Sessa and Nutt 2009).

The State Of Psychiatry Today

In psychiatry today we are at a point where general medicine was in the late 19th century. Smallpox, systemic infections and respiratory disorders were the killers. The 19th century physicians could adequately categorize and classify the problems – they knew who got these infectious diseases and how to diagnose them – but effective and agreed treatments were beyond their grasp. The antibiotic drugs were just around the corner, and once discovered mortality statistics fell dramatically. Psychiatry today is in a similar state. We know who gets depression, psychosis, anxiety and addictions. We are obsessed with the epidemiology of these conditions and write detailed diagnostic manuals to help us accurately identify them. But we lack effective agreed treatments for these common problems. Where in psychiatry is our antibiotic?

Why Traditional Post-Trauma Psychotherapy Doesn't Always Work For Post Traumatic Stress Disorder

The diagnostic label Post Traumatic Stress Disorder (PTSD) has expanded in recent years. Where the diagnosis used to apply primarily to a single, catastrophic, life-threatening event such as a car crash, natural disaster or terrorist attack, the label is now appropriately being used when a person develops the classic cluster of symptoms associated with a less immediately life-threatening, but equally psychologically traumatizing, event such as repeated episodes of child abuse (Roth et al 1997). PTSD, from whatever the cause, tends to result in painful daytime flashbacks and nightmares in which the sufferer experiences a terrifying revisiting of their traumatic memories, triggered by environmental cues. The mainstay of psychological treatment involves gentle re-exposure to these traumatic memories alongside relaxation exercises and learnt coping strategies. Along with the partially useful effects of drugs such as antidepressants, anxiolytics, antipsychotics and mood stabilizers, this 'exposure therapy' is only effective for around 50% of PTSD sufferers (Schottenbauer 2008). But for the remaining patients, attempts at re-visiting painful memories results in dissociative episodes, disengagement from therapy and a spiral into alcohol and drug misuse (Jakupcak 2010). Despite all the best available treatments PTSD remains untreated in half of all sufferers.

The Therapeutic Mechanism Of MDMA

MDMA has a unique pharmaco-psychological profile that makes it perfectly suited for post-trauma psychotherapy. A potent serotoninergic agonist at 5- HT1A and 5-HT1B receptors, MDMA causes euphoria and anxiolysis, reduced fear response and anti-depressant effects (Brunner and Hen 1997, Graeff et al 1996). MDMA's activity at 5-HT2A receptors, the primary site for the classical hallucinogens LSD, DMT, mescaline and psilocybin, results in new ways of thinking about old problems (Nash et al 1994, Liechti and Vollenweider, 2000). Furthermore, release of dopamine and noradrenaline raises levels of arousal and awareness, increases the sense of readiness and improves recall of memories of stressful events (Cozzi et al 1999, Fitzgerald and Reid 1990). And at alpha2-receptors a simultaneous relaxation and reduction in hypervigilance puts the MDMA-user into the Optimum Arousal Zone (Lavelle et al 1999, Foa et al 2009). MDMA also causes a release of oxytocin from the hypothalamus, producing an improved bonding effect and raised levels of empathy (Thompson 2007). A recent study comparing MDMA against intranasal oxytocin demonstrated the former produced greater improvements in pro-social communication than the latter (Kirkpatrick 2014a). In combination these multiple and varied effects give rise to MDMA's description as an 'empathogen' or 'entactogen' (Cami et al 2000). It is the perfect candidate for facilitating psychotherapy – especially for patients with post-traumatic symptoms – as it can help a traumatized patient reach a position of empathic understanding and compassionate regard; an important part of the resolution and remittance of their PTSD symptoms (Sessa 2012).

How MDMA's Characteristics Facilitate Psychotherapy

A recent study looking at the mechanisms behind MDMA's use as a tool for trauma-focused psychotherapy show that participants given MDMA are more likely to use words relating to friendship, support and intimacy, in comparison to the drug methamphetamine, which by contrast reduced participants' discussions about compassion (Bedi 2014). Another study used a simulated experimental paradigm of social exclusion to demonstrate how participants taking MDMA exhibited reduced social exclusion phenomena and enhanced quality of social interactions (Frye 2014). Similarly, MDMA has been shown to enhance levels of shared empathy and pro-social behavior compared to placebo (Hysek 2013). And MDMA can facilitate a faster detection of happy faces, and reduce the detection of negative facial expressions, which leads participants to view their social interaction partner as more caring (Wardle 2014). Furthermore, the pro-social effects of MDMA appear consistent across different experimental environments; Kirkpatrick examined subjects in San Francisco, Chicago and Basel and found the pro-social effects of MDMA are broadly similar (Kirkpatrick 2014), which is an important phenomenon when considering how the drug could be developed widely as a tool for psychotherapy.

Recent Clinical Studies With MDMA

In 2010 a proof of concept placebo-controlled pilot study of MDMA-Assisted psychotherapy for PTSD was published. It showed that 85% of a population of treatment-resistant PTSD sufferers in the USA no longer met the diagnostic criteria for PTSD after a single course of MDMA-assisted psychotherapy, compared to 15% of those receiving placebo. The course consisted of two or three MDMA-assisted sessions given as part of 16 weeks of weekly therapy sessions (Mithoefer 2010). The results were sustained at long-term follow-up of 3.5 years, with no further MDMA interventions required and many patients reducing or stopping their other psychiatric medications (Mithoefer 2013). A similar Swiss MDMA Psychotherapy study also demonstrated substantial improvements for treatment resistant PTSD (Oehen 2012, Oehen and Chabrol 2013).

Why The Pharmaceutical Industry Does Not Like Psychedelic Therapy.

We have a problem in the field of medical research. Development of any new drug for clinical practice costs around fifty million dollars and takes up to fifteen years. The big pharma companies are only prepared to put this level of capital spending in if they can produce a drug at the end of the process that will sell widely and be used in the long-term by patients. The current raft of anti-depressants, anxiolytics, mood stabilizers and anti-psychotics fit this bill perfectly; they are drugs that need to be taken day-in-day-out for years. But psychedelic therapy requires only two or three doses of MDMA, psilocybin or LSD. There is little money to be made from repeated, long-term dosing. Furthermore, all the well-researched psychedelic drugs are off patent. No drug company owns the rights to the substances so anyone with an appropriate license can synthesize them. With no great profits to be made who on earth is going to put all those tens of millions of dollars into the research and development required to see psychedelic drugs licensed? Certainly not the pharma companies. But not only because they cannot reap the rewards of selling the drugs long term. There may be another, more significant and sinister reason why big pharma don't like psychedelics: it may be simply because they work. A single effective course of psychedelic therapy could bring about lasting remission of symptoms, allowing patients to come off their other daily psychiatric medications and recover from their chronic psychiatric problem. Believe or not, in the context of chronic mental disorders it may be that 'cure' is not what pharma want to see.

The Future For Psychotherapy

I would not go as far as to say psychiatrists may look back 20 years from now and wonder why we ever did psychotherapy without drug assistance. That would be going to far. But there is certainly a lot to be said for psychotherapy becoming more focused, more targeted and more intensely delivered through the use of concomitant adjunctive acute psychotropic drugs for some cases – especially those cases where the severity of the disorder itself prevents the delivery of effective psychotherapy. But psychotherapists and psychologists – professions who have sometimes been uncomfortable bed-partners with the use of any psychiatric drugs - will need to change their ways. The old adage that "psychotherapy can only be done with the sober patient" needs to be challenged. Why not utilise the power of a drug like MDMA to help shift the psychotherapeutic process in a certain direction? The main value of the treatment still lies in the therapeutic relationship between patient and therapists as it always has, and indeed with MDMA Therapy most of the sessions between patient and therapist do not use MDMA. But why not take advantage of what MDMA has to offer for two or three of the 16-week sessions? Methods for conducting psychedelic therapy – honed over the last 60 years - are still yet to be fine-tuned. There are some important and necessary challenges to the traditional model of a male-female co-therapist pair. New treatments need to be realistic and clinically deliverable so that they can be rolled-out into large populations. Brief therapies and mini-treatments; these all need to be considered. A lot of valuable knowledge has been neglected and lost since the demise of psychedelic therapy at the end of the 1960s. This is a subject in re-birth with a lot still to learn.

The Future For Medicine

With widespread gene therapy research increasing the future of medicine lies in individualized treatments. In this context, when it comes to psychotherapy, one-size psychological therapy, CBT for instance, does not fit all patients. Nor does the idea of blanket-bombing mental disorders with blunt instruments like SSRIs, mood stabilizing or antipsychotic drugs. This is where psychedelic drugs fit in; they offer bespoke, targeted psychotherapy that gets to the heart of the individual's personal psychological issues. The future of medicine will be driven not by faceless pharma companies but by patient-user power. People want holistic, natural and personalized medical care. This could be the perfect environment for psychedelic therapy to emerge. Our attitude to psychotropic drugs has become more mature and sophisticated in recent years. The virginal approach to mind-expansion that characterized (and killed) LSD's mass discovery in the 1960s is now water under the bridge. Today there is less novelty about psychedelic drugs and (sadly, perhaps) less emphasis from scientists on using these substances to change the world. The general public are also savvier about drugs than they were forty years ago. The public can see the clear distinctions between those drugs that create ghettos, cause dependence and restrict living opportunities and those drugs that are aligned with safe recreational use by artists, musicians, philosophers or anyone seeking to explore their existential place in the world. The killers are crystal meth, crack cocaine and alcohol. Not LSD, psilocybin, cannabis and MDMA. Drugs have moved on. People have moved on. It is time for psychiatry to do the same.

End Note

The godfather of psychoanalysis, Sigmund Freud, had little to say about psychedelic drugs. But he was considerably more aligned with neuroscience than his contemporary Jung (who interestingly didn't approve of psychedelic drugs as tools for self-actualization, despite the subsequent widespread embracing of his ideas by the post-60s psychedelic generation). As a neurologist by training, Freud, had he been around to see the work of psycholytic therapy pioneers such as Dr Ronald Sandison in the UK in the 1950s, would, I believe, have approved of the combining of his beloved psychoanalysis with the drug LSD. Freud believed rightly that changes in brain chemistry underpin mental processes and he made rudimentary statements that were ahead of his time in neuroscience. I leave this essay with a quote from Freud made in 1938, the year he died and the year Albert Hofmann made the first, as yet undiscovered, synthesis of LSD-25: "The future may teach us how to exercise a direct influence, by means of particular chemical substances, upon ...the neural apparatus. It may be that there are other still undreamt of possibilities of therapy." From An Outline of Psychoanalysis. Dr Sigmund Freud. London 1938.

Psycheology: The Study Of The Soul
Neal M. Goldsmith, Ph.D.

Excerpted and revised from: Goldsmith, N.M. Psychedelic Healing: The Promise of Entheogens for Psychotherapy and Spiritual Development. Rochester, VT: Healing Arts Press, © 2011. Printed with permission from the publisher Inner Traditions International. www.InnerTraditions.com

Who owns the mind? Is it the believers in spirit, that illusive "thing" that isn't a thing, but somehow resides in the brain . . . or is it the heart? Do scientists own the mind? Those dissectors and understanders who deny something just because they haven't seen it yet? Before Wilhelm Wundt opened the first experimental psychology laboratory in 1879, there was no academic discipline of psychology separate from philosophy or biology. Perhaps it should have stayed like that – for a while longer at least: the study of mind from a physiological perspective as a subfield in biology and the study of mind from a conceptual perspective as a subfield of philosophy.

Although there are more psychological issues today that can be significantly and reliably treated by a particular psychological approach than there were one hundred years ago, it remains the case that for most psychological complaints, schools of thought or academic orientation are not related to successful treatment. Rather, it is similarity of background and values and the creation of a trusting rapport that are most correlated with successful psychotherapy. Furthermore, for common "neurosis," talk therapy with a skilled practitioner (or even a trusted family member) is more effective over the long run than an equivalent-length treatment with any pharmaceutical alone. Psychotherapy combined with the *short-*

term use of an appropriate pharmaceutical is the most effective treatment, but the pharmaceutical usage *must* be proscribed. This is because pharmaceuticals tend to backfire after prolonged use—backfire due to tolerance and side effects, where the benefit begins to be outweighed by the drawbacks. The current tendency to prescribe a pharmaceutical, simply because it works at first, is mistaken. We must find combinations of treatments that are explicitly chosen to be effective without relapse when the chemical is finally withdrawn.

Psycheology: The Study of the Soul

This article integrates my own evolution into the discussion of using psychedelics for healing. I can illustrate this point by defining a word I've crafted and like to use in my practice, the word *psycheology*. You won't find the word *psycheology* in any dictionary (I've searched). Rather, it is a made-up word—a neologism (from the Greek: *neo* meaning "new" and *logos* meaning "word" or "statement," or meaningful sound – information as patterned energy). *Psycheology* is a word I created in my effort to reclaim the true, original meaning of the word *psychology*.

The word *psychology* comes from the Greek *psukhe,* meaning "soul," "spirit," "mind," "life," and "breath," combined with the Greek *logos,* here used as "statement," "expression," and "discourse," more often thought of today in the form of "-ology," as "the study of." Although the academic and clinical discipline of psychology has become a medical—and therefore a pathology-oriented—field, prior to the late 1800s, the study of our inner mental life was the study of our soul, our deepest self or essence.

My purpose in reclaiming the word soul for personality theory is not supernatural, religious, or even spiritual, but rather, to bring psychologists, my clients, and us all back to psychology as the study of the psyche, to a focus on the ground of our being, to the soul, because it is this part of us that is the earliest, deepest, and the most authentic part of us. From a psychotherapeutic perspective, the psyche is the part of us that is the most influential in effecting behavioral change and improving self-esteem. Not coincidentally, it is also the part of us that we see illuminated during the psychedelic experience, and it is this illumination of our true nature (or the corresponding "death" of our identification with the ego) that accounts for the therapeutic value of the psychedelic experience. This effect is similar to the concept of sympathetic vibration; wherein a still tuning fork brought into contact with a vibrating one will begin to vibrate at the same frequency. If our conscious attention or identity is brought into contact with or awareness of our deepest ground of being, our conscious awareness elicits or comes into identity with—becomes—that same deepest sense of self. We are changed—transformed— back into identity with the true self we abandoned in our childhood quest for parental love.

To foster this process of re-identification, we must come to view much of behavior now labeled "neurotic" not as pathological, but as the organism's natural response to developmental and environmental stresses on the path to maturation. From this perspective, "neurosis" is better seen as developmental challenge—the surmounting of which brings maturity or wisdom—rather than as pathology.

The term neurosis, as generally applied, is not accurate or helpful. In fact, one of the most negative influences on mental health is the "sick" concept itself, which tightens and distorts, keeping us from a natural unfolding and realignment.

In essence, we need to return the clinical practice of psychology to the unfolding of the psyche, in all its beauty and complexity, as a non-medical, natural phenomenon.

With the exception of biologically-based illnesses, psychology must come to be seen as the science of spiritual maturity. We call people "neurotic" when, in reality, it's not a medical illness they are suffering from, but spiritual immaturity. We must redefine spirituality, too, not as supernatural, but as simply the natural unfolding toward the wise, mature end of the normal curve of human developmental psychology.

In my practice, I find over and over again that big-picture understanding, active listening, and fundamental positive regard work best. From my perspective, "healing" takes place only when we get underneath our modern imago, persona, or personality, to rest at the ground of our being—to naturally unfold according to our perfect, inner template for development. That process both requires and facilitates the emergence of self-acceptance and will.

The Psycheology Approach to Psychotherapy

To summarize, in their approach to clients, therapists with the psycheology worldview will tend to naturally express many approaches from the following list of philosophies and methods:

• Psycheology is about the direct experience of the foundation of our true self. I want to emphasize that in psycheology, we are not talking about the personality, but the true, original self—the self we were born as, before our parents "had at us." Our true, original self lies under our personality, in the transpersonal ground of our being, at our core.

• As newborns, we are all perfect. Of course we all have individual differences at birth, like the wide-ranging forms of trees in the forest, yet we are all "perfect" in our essence.

• Love—safety—is the central issue of infancy; lack thereof results in defensive highjacking of the ego function and the creation of a personality as a strategy to attain love.

• Personality is a strategy devised by an earlier, immature version of our adult self and does not have the complexity or sophistication to support a healthy adult life.

• Neurosis is the natural, stepwise unfolding of human maturation. It's not about pathology, but spiritual immaturity.

• The desire for change is a reflection of the problem, not of the solution. So, working on yourself or your relationships doesn't work. Rather, the only thing to "do" is simply to be; and simply being is not the result of an active pursuit, but rather the natural result of releasing the self from the encumbrance or distraction of an immature personality strategy.

• Transformative developmental change is possible through a stepwise, dualistic dance—a combination of transcendent change that touches soul and reaches forward, and cathartic change that removes unconscious chains and releases the past.

• When we awaken to our projections, we feel empathy and acceptance—love—for our parents and ourselves. This enables us to relax and release the knot in our psyche, to dis-identify with the acquired, defensive shell of personality and re-identify with our true core, our original self, through a process of retroactive maturation—to finally complete our childhood—rejoining our true psyche in the unfolding of identity that was stunted as a baby. This entails a process of repositioning the locus/focus of our awareness—of self—to the ground of our being, to the truest self, to the psyche.

• The psycheology approach is yogic, seeing human development as a perfect, healthy, developmental process of unfolding maturation. Psycheology approaches the human organism as a single whole seen sometimes as body, sometimes as mind, sometimes as spirit, but most effectively approached as the integral of all three.

• Psychedelic therapy can be a safe and extremely effective tool in facilitating transformative developmental change by enabling us to see ourselves with love and to safely engage in catharsis. Stunted or skewed development can be gotten back on track, but psychedelics are not cognitive development—or enlightenment—in a pill. Psychedelics can trigger insight, but behavior change takes time, and in this culture, such realignment is often harder to sustain than we acknowledge.

• Effective methods exist for changing policies and bureaucracies, and we are honor bound to bravely apply them in the pursuit of science, truth, and freedom.

• Having seen these key lessons, as good global citizens, we are compelled to actively apply these finding, to improve the world.

• It's important, too, for us to consider the future of psychedelic therapy and policy – and how the re-integration of psychedelics into western civilization can provide a rite of passage for our culture as a whole, healing Cartesian dualism, and elevating us to a new, integral level of society.

Finally, I would be remiss if I didn't emphasize again, the effectiveness of meditation, and similar/related practices, in the achievement of the kind of grounded, peaceful, happy stability that we are exploring here. I believe any time-tested, sincerely embraced practice would serve to mature us, but psychedelics have the added quality of revealing our true self to ourselves fully, as if for the first time. Seeing through anger, fear, judgment and defensiveness, to our loving, loveable core is a hallmark of a positive psychedelic experience—and of the psycheology approach.

Oriental Jones And The Medal Of Freedom

Mike Crowley – *July 2014*

William Jones Jnr. (1746-1794) was the son of the man who invented π.[1] Young William had an insatiable thirst for knowledge and once said, "I hold every day lost, when I do not acquire some new knowledge of man and nature." He was an intellectual prodigy in many respects[2] but the field in which he really excelled was languages. By the age of twenty Jones had mastered French, Italian, Spanish, Portuguese, Latin, Greek, Persian and Arabic. King Christian VII of Denmark heard tell of this outstanding young linguist during a visit to London. The king happened to have a Persian text with him and he requested that Jones translate it from Farsi into French.[3] The success of this translation led to Jones' admission into Dr. Samuel Johnson's elite "Literary Club" and to the scholarly Royal Society, where he first met Benjamin Franklin.

While sitting as a judge in the Welsh circuit courts, Jones published Poems, Chiefly Translations from Asiatick Languages, together with Two Essays on the Poetry of Eastern Nations and on the Arts commonly called Imitative (1772). It is hard for us today to imagine any book with such a ponderous title selling more than a handful of copies but, by the standards of its day, it was a best-seller. This innovative volume earned him the nickname "Oriental Jones" and inspired a clique of young drug-taking poets to investigate Asian ideas and esthetics. When these writers, Coleridge, Wordsworth, et al., published their own works they kicked off something called the Romantic Movement. In fact, the cultural fallout of Jones' activities was still evident in the late 20th century - Hollywood's "Indiana Jones" was inspired by William "Oriental" Jones.

An outspoken radical, Jones opposed the slave trade and the new war with the American colonists. These colonial sympathies and his close personal friendship with Ben Franklin, led the British government to use him as an intermediary in peace negotiations. In May of 1779, he was sent to Paris where he attempted to broker a solution to the American conflict by presenting Franklin with the British peace offer, thinly disguised as ancient text about trouble between Athens and one of its colonies. Like all well-educated men of his day, Franklin was familiar with Latin and Greek but he could not accept the terms implied in the document. Although Franklin rejected the British offer, this did not diminish his friendship with Jones as we see from their subsequent correspondence.

In 1783, Jones received a letter from Franklin telling him about the *Libertas Americana* ("American freedom") medal. Franklin's letter implies that he and Jones discussed the proposed medal while in Paris. Franklin wrote, "The engraving of my medal, which you know was projected before the peace, is but just finished... You will see that I have profited by some of your ideas, and adopted the mottoes you were so kind as to furnish." The obverse of this medal shows the head of "Miss Liberty" against a background of a liberty cap on a pole. The red, woolen *pileus* ("liberty cap"), a cap worn by freed slaves in ancient Rome, was a popular symbol for "liberty" and in the colonies a cap on a pole signified defiance of the British. Another meaning of *pileus* was introduced in the middle of the 18th century: it is the scientific term for the cap of a mushroom. As members of the Royal Society, the premier scientific body of their time, both Franklin and Jones would have been exposed to the latest in technical terminology. Another term introduced about the same time as *pileus* was *stipe*, the botanical term for a mushroom's stalk or stem. The original meaning of the Latin *stipe* was "pole" or "stake," so we may assume that Jones, fluent in Latin, would make the connection between the American symbol of cap-on-pole with *pileus*-on-*stipe*. Furthermore, as an active member of the Royal Society, it would be surprising if he did not connect this with the recently coined terminology of mushroom anatomy.

The proportions of Franklin's medal and its beautifully executed bust of Liberty have made this one of the most sought-after coins in the world. However, in contrast to his naturalistic depiction of Liberty, the cap-on-a-pole that leans diagonally behind her is extremely stylized. The cap, in particular, is unlike any other

representation of this symbolic headgear and is unlike the coinage subsequently based on the medal. Rather than falling in limp folds, as cloth should [see 1794 half-cent, below], it is smooth, rigid and symmetrical about the pole. To be frank, it bears an uncanny resemblance to the "liberty cap" (*Psilocybe semilanceata*) mushroom, not just in its shape and the proportions of cap and stem but also in that it appears to mimic the mushroom's "acute umbo" and "striated margin," both of which are distinguishing features of this species.

A detail of the Libertas medal (left) shows the stiff, stylized version of the "liberty cap on liberty pole" motif. The right image shows the same detail overlaid with a photograph of an actual "liberty cap" mushroom.

All mushrooms of the *Psilocybe* genus have an umbo, a small bump at the center of the cap but *P. semilanceata* is notable for its particularly pointed bump (an acute umbo). The cap of

a *P. semilanceata* mushroom also has a translucent band around its outermost edge allowing its gills to be visible as a band of vertical stripes. This is called a "translucent-striate margin."[4] Though small, this "highly to extremely potent"[5] entheogen forms extensive colonies in meadows of rye grass and grows in great profusion on the green hills of Wales, Jones' homeland. Small brown mushrooms tend not to be given common names, especially in mycophobic lands such as Great Britain, and the name "liberty cap" seems to be a scholarly invention rather than a truly "common" name. It is not clear when this insignificant species was first named the "liberty cap" but it has borne this name since at least 1841.[6]

Later that year, at the age of 38, Mr. Jones was knighted "Sir William" and posted to Fort William (modern Calcutta) as

Half-cent of 1794
with a realistic cloth

Chief Justice of Britain's new colony, India. Despite his onerous judicial duties, Sir William found time for writing and scholarly research. A year after arriving in India (1784) he founded the Asiatick Society of Bengal, the first Western organization dedicated to studying a foreign culture. He presented his own findings in annual addresses to the Society and it during his 1786 discourse he opined that the Sanskrit and Persian languages are related to Latin, Greek and, somewhat more distantly, to "Celtick" and "Gothick."[7] He concluded that these languages all derive from a common ancestor, now lost.

Jones delighted in India's plethora of languages and dialects, religions and ethnicities. He lived in India for little more than a decade until his death in 1794. In addition to his linguistic research he enthusiastically investigated, and wrote dissertations on, all facets of Indian culture – from elevated topics such as its religions and methods of reckoning time to more mundane details of its food, music and board games. At the end of his life he admitted to knowing thirteen languages well, and having "a moderate acquaintance" with twenty-eight others.

In a paper on Mithraic survivals in the Masonic tradition of the late 18th and early 19th centuries,[8] Mark Hoffman and Carl Ruck draw a connection between psychoactive mushrooms and the "liberty cap on liberty pole" motif. While their conjecture chiefly concerns the *Amanita muscaria* mushroom, they do make a passing mention of *Psilocybe semilanceata* and assert that an allegorical image in a Masonic work of the late

19th century conceals an image of *P. semilanceata*. However, it should be pointed out that the alleged image displays none of the characteristic features by which this mushroom may be identified and their "unusual and at first unidentifiable object... perhaps a lantern" is quite clearly a depiction of an open book.

It is generally believed that, until the 1950s, the psychoactive properties of *Psilocybe* spp. were known only to a few Mazatec Indians. However, I can recall an encounter with a Welsh child in the mid-1950s who declared that he ate "toadstools" because they let him "see the fairies" and, as stated above, *P. semilanceata* is very abundant in Wales. Given the omnivorous habits of small children it is unlikely that the properties of this small mushroom could have gone entirely unnoticed. One cannot ignore the possibility that a tradition of *P. semilanceata* use may have existed, perhaps as a secret kept by "wise women," "cunning men" and hedgerow herbalists. Surely, the eternally inquisitive Jones would have sought out just these kinds of people for information about his environment while living in Wales.

It is apparent from Franklin's letter that, quite apart from the mottoes, the *Libertas Americana* medal incorporates other "ideas" contributed by Jones. I, for one, would love to know what some of those "ideas" were.

The 4 Archetypes Of Psilocybin
James W. Jesso - *May 2014*

The Story So Far

In 2010, I began a thirteen-month journey of utilizing psilocybin mushrooms to heal a (self-diagnosed) depression and borderline psychosis. This illness was the result of a year of reckless lifestyle choices filled with substance abuse and happening on the other side of the planet, far from my normal support structures.

I was living in an ongoing party, ignorant to being 'trapped' until the inevitable damage of my poor choices laid their pernicious blows upon my psyche. After a year of this lifestyle, I eventually recognized myself as sick and knew that I needed help to emerge from the self-deprecating wallow I was festering in.

In some strange concoction of validations, I decided that psilocybin mushrooms could help me. So once a month, every month, I took an upper-medium dose of P. cubensis and went alone into the night, into nature, and asked to be shown my wounds and prayed for resolution.

There was pain, pleasure, crying, laughter, cosmic loneliness, and a seeming connection with a dimension of unconditional loving support. I navigated the darkest, but also the most vibrant recesses of my self. I sat repeatedly through tremendous outpourings of potent emotions, cascading from my unconscious in a cathartic process of finding that which I hadn't realized I had long lost: confidence in being worthy of love.

When this thirteen-month healing journey began, I was living in my parents' basement, angry at the world, unhappy with my situation, feeling horribly alone, regretful, and shameful. I was going out of my mind with the feeling that I had no one I could really talk with about my situation; I was suffering the tragic self-perpetuating abyss of perceivable isolation from support that riddles the depressed mind.

At the end of these thirteen months, I felt as though I was thriving. Living on the other side of the country, I was involved with a vibrant community, had a stable job, was working towards a healthier diet, perceivably connected to my creative passions, happy, and comparatively resolved in the challenges I had entered a relationship with the mushroom to address.

In 2011, I began writing down what I'd learned during this transformational period in my life. I mapped a construct for how I navigated the psilocybin experience towards the potentiation and successful integration of psychospiritual maturation and personal healing. The hope was that sharing my story and what I learned might help others lost in similar situations.

In 2013, with the help of a crowdfunding community, I published my first book, Decomposing The Shadow: Lessons From The Psilocybin Mushroom. It presents a conceptual model for the psilocybin experience as it pertains to psychospiritual transformation and mental-emotional healing. That same summer, I travelled across Canada, teaching nineteen events (readings and lectures) in four and a half months. The lecture I taught was called The 4 Archetypes Of Psilocybin, a restructured presentation of the model presented in Decomposing The Shadow.

Since then, I have continued to expand and explore my philosophies through multiple essays, articles, podcast interviews, public presentations, an audiobook, and a book titled Soundscapes & Psychedelics (2014).

In this article, I will present the basic foundation of Decomposing The Shadow as it was presented in my lecture series. It will also include concepts presented in later works, as well as updated perspectives based on recent research with psilocybin. This article will present an overview of my most updated construction of the 4 Archetypes model, as of the time of its writing.

Much of the information presented will be overarching statements for which you will find foundational explanations within my other work. It is presented with a rather complex literary style on the awareness that the reader of Psychedelic American does not need larger philosophical concepts to be overtly simplified.

Defining A New Model

That which we understand as reality is a set of experiential content resulting from an energetic feedback system of informational processing. All of reality is inherently meaningless information unconsciously rendered as a meaningful human experience according to the modeling systems, languages, encoded into the psyche. Information enters, gets calculated through the psychosensory nervous system, and is given a meaningfulness according to the internal models used to render the perceived information. That meaningfulness is then attributed to the received data and interacted with as though the meaningfulness unconsciously generated were inherently true, and external. In summary, the models embedded in the psyche define the experiential content of one's reality.

This same process applies in experiences that transcend the normal operating grounds of waking-state consciousness, experiences 'novel' to the normal mind such as psychedelic experiences. The models that we apply to the novel information being processed will influence how it will be experientially rendered. The 4 Archetypes Of Psilocybin is presented as a model that enables the novel information of the psilocybin experience to be rendered and navigated as functionally applicable to one's ongoing process of psychospiritual maturation. Specially, it offers an accessible means to understand, navigate, and integrate the shadow, or emotionally challenging aspects of a psilocybin experience. It presents a logical basis for integrating emotionally dark experiences as potentially the most beneficial to said maturation process, and in doing so, amplifying those experiences' potential benefit and mitigating their potential damage.

That we constantly re-evaluate and refresh models for the psilocybin experience is important, as the current prevailing models hinder a direct accessibility to the full potential of the experience in trying to overstate or suppress the psychospiritual potential. The indigenous models of reverence to visionary plants are, for the most part, culturally inapplicable to the daily life of common Westerners (this doesn't even begin to address the issues of cultural appropriation). The prevailing fringe models, unfortunately, often romanticize the visually symbolic elements of the novel information being offered as a reality of ontological truth (e.g. 'there really are machine elves and jaguars roaming the mind and they are the source of wisdom I am receiving'). The academic models, though advancing towards an integral approach of psychedelics, often present psychedelics to the common person as only 'tools to facilitate scientific research and address psychopathological issues.' This leads to the assumption of psychedelics as merely for correcting psychopathological issues or for institutional research, which cultivates a disconnect from the average user. Finally, the common model for psychedelics within conventional 'big media' culture is that of demonization and/or marginalization, which are not always mutually exclusive in today's "renaissance."

Apart from maybe the latter, each of these models – indigenous, academic, fringe, and conventional – offer potential benefit towards cultivating personal growth. The 4 Archetypes Of Psilocybin gathers the beneficial elements in each of these models, without appropriation, along with all relevant information available within the updated human knowledge base. While maintaining a discerning and healthy skepticism, these elements and related information are compiled into a model founded on both personal experiences and extensive study, and is presented with a contemporary language. The result is a model that enables the effective navigation of the psilocybin experience for the benefits of processing emotional traumas and encouraging mental-emotional wellbeing through facilitating psychospiritual maturation, and can fit within the lifestyle of the common westerner.

This model is not intended as to say "this is how it is," but rather, in essence, "this is what I have found to work; take what works for you, leave what doesn't." None of us have the answers yet; none of us know what the psychedelic experience really is. Yet, as research, theories, and discourse furthers through the psychedelic community and into the public sphere, we further our capacity to ask better questions, and thus we get closer to a fuller understanding of what potential the psychedelic experience can fully offer. This is the intention behind my work, the 4 Archetypes included. What follows for the rest of this article will be a basic rundown of the most updated 4 Archetypes model.

The Mushroom: What Is It? What Does It Do For Us?

Since the experiential content of a meaningful reality (the rendered result of information processed through the psychosensory nervous system) is generated by one's internal models, establishing the basis for 'what' the mushroom is and 'why' one would use it within the currently presented model is of key importance.

If one's intention is to work with psilocybin mushrooms as a means to personal healing of past traumas or for personal development, I believe it is of vital importance that one approaches them with respect. With respect, one can offer the reverence or "regard" of the mushroom as the means to one's intended personal growth.

Furthering the 'what' and the 'why' in how one can "regard" the mushroom within the 4 Archetypes model, the psilocybin mushroom is a living organism that can activate four powerful emotional processes:

- Increased Emotional Potency:
  Turning up the 'fullness' of one's experienced emotions, be it loneliness, sadness, joy, curiosity, grief, excitement, or religious wonder.
- Altered Emotional Meaningfulness:
  The change in emotional potency changes the meaningfulness applied to the perceived dataset of reality, engendering a different quality to its experiential content. These changes engage each other like a feedback loop rendered at one's surface awareness.
- Entrance Into New Environments And Realities:
  This change in meaningfulness and content changes the very experience of one's environment, whether internal or external.

- Catalyzed Emotional Catharsis:

The increased potency and altered meaningfulness of one's emotional experience catalyzes strong emotional currents to be actively expressed and exposed to an individual's surface awareness.

In activating said emotional processes, the mushrooms can be seen as a tool, teacher, or ally that supports a person's healing and growth through cultivating four major results:

- Retroactive Emotional Release:

Enabling access to past emotional repression, allowing those emotions to be processed and properly addressed.

- Dissolution of Mental-Emotional Patterns:

Removing the normal patterns of psychological defenses employed to maintain composure in the face of potent emotions, either light or dark.

- Cultivation of A Broader Self-Awareness:

By bringing both light and dark repressed material to surface awareness, along with its associated memories and patterns, psilocybin offers a fuller picture to the user as to what influences their personality.

- The Cultivation of Psychospiritual Maturity:

The aforementioned emotional processes, along with the stated results, combine to cultivate a dis-identification of negative patterns (strategies for 'defense of' and 'compensation for' embedded traumas) in the psyche, and allow for the natural re-identification with behavioral patterns associated to one's self prior to the strategies instituted into the psyche throughout life.

These emotional processes and psychological results can be correlated to four of the major neurological changes we have observed in brain imaging studies with psilocybin. The first three are based on observations made by Dr. Robin Carhart-Harris in his fMRI research with psilocybin (Carhart-Harris, Robin L., et al., 2012), and the last is based on Dr. Rainer Kraehenmann's imaging research (Kraehenmann, Rainer, et al. 2014).

1.   Through the suppression of blood flow to the posterior cingulate, a major communication hub in the brain, the normally rendered meaningfulness of contextual stimuli, (e.g. 'a tree is just a tree according to my conceptual frameworks for tree') is lessened. This allows for cross-communication within the brain to enable access to meaningfulness associated with deeper psychological patterns (e.g. 'that tree is just like the one I cried under when my childhood dog died').

2.   Increases of sensory activity associated with memories within the brain allow for the deep meaningfulness accessed and the attached memories to be relived at a greater sensorial depth and emotional potency, thus enabling a fuller connection to the emotions associated with those memories.

3.   Through the suppression of blood flow to the medial prefrontal cortex, the area of the brain that associates context to adaptive responses and emotions (i.e. the normal patterns of behavior associated with emotional triggering), psilocybin lessens the strength of trauma-adapted behaviors or patterns. These trauma patterns are typically the cause of most mental-emotional instability. Thus, their dissolution can enable the processing of the heightened emotional content (catharsis) associated with the aroused and sensorially amplified memories, naturally helping to bring about emotional stability.

4. Recent research on psilocybin has suggested that it 'weakens' the amygdala's processing of negative emotions in general. Looking at this research through the 4 Archetypes Model (and from personal experience), the reduction of the amygdala's processing of negative emotions are specific to the negative emotions associated with the defenses surrounding trauma. Thus, the decrease in amygdala function allows for an openness in which challenging emotions (light or dark) can be experienced without the extra load of the "negative emotions" arising in the need for defense from these challenging emotions.

Finally, I will leave you with the four archetypal experiences/lessons within a psychospiritually therapeutic psilocybin experience.

The 4 (Experiential) Archetypes Of Psilocybin

Archetype is a term introduced into the psychology lexicon by the late psychiatrist and visionary Carl Gustav Jung. From what I understand of his work, an archetype is the observable symbolic manifestation of patterns within the psyche that extends beyond the individual self. The symbolic manifestations and the terminology surrounding these manifestations are not the archetypes themselves, which exist only as movements of consciousness. Archetypes inevitably take on symbolic representations as they emerge to surface awareness. The use of archetypes or archetypal processes as a model for understanding the movement of consciousness within the psyche can help in identifying and cultivating an understanding of those movements.

Applying this concept to the psilocybin experience enables the rational categorization, navigation, and integration of these novel movements of consciousness. As I describe these archetypes, keep in mind that they are merely the logical construction of culturally accessible symbols (a model) to support the linguistic genesis of novel movements of consciousness within the psyche. The archetypes are not fundamentally separate sets of experiences, but are conceptually separated in order to better triangulate the subtle combinations in which they emerge within a psilocybin experience. These archetypes function as categories for the types of experiences and potential lessons that psilocybin may offer. Separate only in discussion, they seamlessly dance and merge within the experience itself.

The four archetypes of psilocybin are Surrender, Facing The Shadow, Uncovering The True Self, and Oneness. We will briefly explore each one, leaving their implications undiscussed and up to the reader to cross-examine these archetypes to the previously mentioned processes, results, and neurological changes.

Surrender - The mushroom experience is an emotional experience. One's emotions are always honest (though usually not based in accurate interpretations of context). Thus, the mushroom experience is that of potent emotional honesty. The process of surrender is that of releasing the conceptual control one usually attempts to maintain over the expression of emotional honesty.

These mushroom-catalyzed emotions tend to rush like a waterfall, forcing one into acceptance, surrender, and release: Surrender. This process helps support retroactive emotional release through catharsis. In turn, this can have beneficial effects on one's maturation process through what I call Emotive-Psychosynthesis (explained at depth in Decomposing The Shadow).

The act of Surrender, then, can work as a symbolic practice for learning surrender to emotional honesty in all areas of life, which would in turn allow naturally expressed behavioral patterns to be embodied without the 'negative emotions' attached to repression.

Facing the Shadow - The shadow is the archetypal representation of all the dark or potentially traumatic emotions within the psyche, from which the repression or evasions of most negative behavioral patterns emerge.

To Face The Shadow is to Surrender to the honesty of one's emotional darkness. To allow the light of awareness to shine upon that which we fear acknowledging about ourselves. Doing so allows for the emotive-psychosynthesis of repressed emotions and offers us a broader awareness of 'what' influences 'who' we are and why we act the way we do.

Uncovering the True Self – The True Self is not a defined identity, but a set of observations. In facing one's shadow and surrendering to emotional honesty, we are exposed to the truth that we all possess behaviors bred of unaddressed darkness. In this unearthing of darkness we are offered a window to the True Self as an expression of who we are naturally, without the strategies of evading, defending, and repressing our emotional honesty, both in real-time and within a history of trauma, often with an otherworldly sense of loving acceptance.

The further implications of the True Self can be extended beyond that of the individual, into the metaphysical or religious interpretation of one as being "Truth" itself, The Divine, God, etc. This leads into the last archetype.

Oneness - The awareness and identification with the interconnectivity of all things, be it humanity as one large family, awareness of the integral system of the planet of which we are all a part, or that of more cosmic, non-dual foundations.

Oneness is the archetype that is most difficultly defined, as its essence is ineffable.

Conclusion

For the conscious user, there are several potentially beneficial shifts in behavior and self-identity that can occur through these changes, processes, results, and experiences occasioned by psilocybin. This includes increased capacity for courage, openness to living, creativity, increased perceptions of beauty, personal coherence, self-confidence, and so on. All of these changes are ones observed within myself as a result of how the mushroom experiences I underwent changed the manner in which I engaged my life.

In exploring the implications of this model and psilocybin in general, I feel there are deeper discussions to be had about what the mushrooms show us about the nature of emotions, how we process the socially conditioned models for reality embedded in the psyche, the gains and losses of our current conventional psychological models, the neurobiological implications of experiencing spiritual wonder and identifying with Oneness, how all these elements relate to health of the mind and the physical body; even the potential mutability of reality as an experience in general, and many more wonderful notions of greater magnitude. The ones that I continue to be excitedly engaged in are always being explored through my writing and research.

May the discourse on psychedelics continue to be expanded, refined, and progressed, allowing it to continue enlighten and inspire us.

Cannabis and Alzheimer's Disease: Prevention and Treatment

Clint Werner

Alzheimer's disease (AD) is a neurodegenerative syndrome that develops and progresses undetected over a number of years, although impairment often escalates rapidly once symptoms become evident. Alzheimer's disease symptoms usually first manifest as annoying forgetfulness, then progress through confusion and disorientation to a terminal, near-vegetative state. It is believed that the pathology is initiated many years before impairment is evident which provides a large window for attenuating the damage with therapeutic agents. Interestingly, the phytocannabinoids found in the cannabis plant have shown promising activity against many of the destructive influences involved in the complex pathology of AD.

Each of our brains is made up of about 100 billion neurons which communicate through vast neural networks. Alzheimer's disease is a progressive and fatal brain disease that destroys neural networks by apparently altering a protein which becomes toxic and forms plaque deposits outside of cells while destroying nerve fibers inside of cells and leaving them in tangles. These plaques and tangles were first described in the medical literature in 1906 by Dr. Alois Alzheimer who found them during an autopsy on a severely-demented patient.

We now know that an enzyme alters the production of amyloid-beta protein, possibly in nature or volume or both, and excessive levels of this protein seem to create problems inside and outside of neurons. The exact progression and interaction of influences on AD is still unknown but it appears that amyloid beta has a toxic effect on the structure of microtubules which are involved in the relay synaptic transmissions. Tau is a protein necessary for the formation and integrity of the microtubules. It is thought that excessive levels of amyloid beta protein increase levels of an enzyme, glucose synthase kinase 3 (GSK3) which alters tau protein causing it to migrate away from microtubules.[1] Without tau, the structures collapse and twist into the neurofibrillary tangles identified by Dr. Alzheimer. The amyloid-beta (AB) fragments also gather into plaques which surround neurons and disrupt synaptic transmission and generate inflammation. Both plaques and tangles characterize Alzheimer's disease; tau tangles without plaques occur in other forms of dementia but AD pathology involves the presence of plaques.

This complex cascade of damage is accelerated when the trauma and inflammation over-stimulate the brain's immune cells, microglia, which then begin to harm healthy neurons. This damage then produces more toxic compounds and results in even more inflammation, which further drives the neurodegenerative process.

An effective drug for treating or preventing AD would, ideally, have activity against all of these influences—production of toxic AB fragments, clumping of AB into plaques, increased production of GSK3, alterations of tau protein in microtubules which lead to neurofibrillary tangles, and over-activation of microglia cells. Interestingly, the cannabis plant seems to have beneficial activities in all of those areas.

In 1988 the first cannabinoid receptor (CB1) was located and subsequent mapping of the receptors' distribution revealed them to be the most prominent G-coupled protein receptors in the central nervous system. Investigators studying the functions of the receptors discovered that they are involved in neuroprotection.[2] A National Institutes of Mental Health-sponsored study confirmed that the natural cannabinoids THC and CBD have the ability to effectively reduce toxic levels of glutamate and protect neurons from death by acting as antioxidants. The researchers discovered that THC and CBD were superior to other antioxidants in preventing oxidative cell damage and that "CBD was more protective against glutamate neurotoxicity than either ascorbate [vitamin C] or alpha-tocopherol [vitamin E], indicating it to be a potent antioxidant."[3] Research published in 2008 reported that the cannabinoid, cannabidiol (CBD) "exerts a combination of neuroprotective, anti-oxidative and anti-apoptotic effects against beta-amyloid toxicity."[4]

A group of investigators found that the cannabinoid THC effectively decreased levels of an enzyme necessary for the generation and deposit of toxic AB into plaque lesions. The effect was so pronounced that they stated that "Compared to currently approved drugs prescribed for the treatment of Alzheimer's Disease, THC is a considerably superior inhibitor of Abeta aggregation, and this study provides a previously unrecognized molecular mechanism through which cannabinoid molecules may directly impact the progression of this debilitating disease.[5] Studies in animal models confirm the findings in cell cultures. Prolonged oral treatment with a synthetic cannabinoid, which acts in a manner similar to THC, produced significant beneficial effects in mice with induced AD symptoms by decreasing inflammation and clearing AB from the brain.[6] Another study confirmed that THC effectively decreased AB plaque and neurodegeneration in AD animals.[7]

Cannabinoids also work against the changes in tau proteins that lead to neurofibrillary tangles within brain cells. A study with a blend of natural plant cannabinoids found that they quickly and significantly reduced the markers for AD neurodegeneration and improved behavior and cognition in research animals.[8]

We also see evidence that there is a lack of CB1 expression in the brains of AD patients and that activating both CB1 and CB2 with cannabinoids "blocked beta-induced activation of cultured microglial cells" and stemmed the neurotoxic reaction from beta-amyloid activated microglia. It was proposed that "cannabinoids succeed in preventing the neurodegenerative process occurring in the disease."[9] The data clearly indicates that cannabinoids play a critical role in the avoidance or development of AD pathology. A study in rats determined that the damage caused by AB injections could be decreased by preserving the levels of the body's naturally-occurring cannabinoids, which resulted in memory retention.[10]

A study published in 2014 determined that THC restricts the production of a damaging inflammatory agent that promotes AD, lowers the levels of AB, prevents AB clumping into plaques, and enhances the vitality and energy production of the neurons. The results were so impressive that the researchers stated, "These sets of data strongly suggest that THC could be a potential therapeutic treatment option for Alzheimer's disease through multiple functions and pathways."[11]

Yet, the neuro-protective activity of cannabinoids is not limited to the CB receptors and seems to involve an array of other receptors. The NMDA receptor also protects the brain from neurotoxicity and interacts with certain cannabinoids to decrease brain-cell death.[12] Evidence is accruing that cannabinoids influence a number of other receptor systems to produce their beneficial effects.[13]

A new perspective on disease is emerging as we learn more about the role that chronic inflammation plays in initiating degenerative changes that manifest as seemingly different but actually related syndromes, including AD, cancer, diabetes and others. Harmful inflammation plays an integral role in inducing and promoting AD.[14] Cannabinoids are powerful anti-inflammatory agents that work by reducing the production of inflammatory compounds, modulating immune response and enhancing mitochondria function.[15] A large population study found that subjects who smoked marijuana a least three times per week had less than half the levels of the inflammatory marker C-reactive protein in their blood compared to nonusers.[16] This decrease in inflammation, in part, explains how cannabis can have beneficial effects on such a wide array of diseases.

Unfortunately, despite the broad margin of safety that characterizes cannabinoid use, uniquely confounding restrictions have limited the ability of researchers to investigate cannabis in human subjects. There is an appalling dearth of clinical research given the remarkable range of neuro-protective effects and safety from cannabis. Alzheimer's disease is a devastating and enormously expensive disease to cope with. As far back as 1998, the authors of a paper on the future impact of AD projected that "If interventions could delay the onset of the disease by 2 years, after 50 years, there would be nearly 2 million fewer cases than projected; if onset could be delayed by 1 year, there would be nearly 800,000 fewer cases" and that this would "have a major public health impact."[17]

Limited clinical research has been carried out with cannabinoids for AD patients. A few human trials with the synthetic cannabinoid, dronabinol found that AD patients were less agitated and gained weight while using the drug.[18] Pharmaceutical companies continue to spend millions of dollars seeking a drug that can effectively treat AD while ignoring or dismissing the promise of cannabis as a neuro-protective agent. Hopefully as the social and political forces begin to change, cannabis can be studied far more extensively in human subjects.

The Evidence for the Effective Use of Cannabis Extracts in Treatment of Various Cancers

Justin Kander

The use of cannabis to treat the side effects of chemotherapy has been widespread for decades. It is a generally accepted fact that smoking cannabis helps alleviate nausea and makes cancer patients feels better. However, the truly controversial issue lies in using cannabis to directly treat cancer and put it into remission. While there are no formal clinical trials that have outright proven human anti-cancer effects, the existing scientific and anecdotal evidence is truly overwhelming. This body of evidence leaves no doubt that cannabis extracts can fight cancer in humans.

The scientific studies alone lend powerful support to the theory that cannabinoids could treat cancer in humans. First, cannabinoids have been proven to directly induce programmed cell death in many types of common cancer cells. Two studies by Dr. Manuel Guzmán showed THC induced apoptosis in brain and pancreatic cancer cells through[1,2]. THC can also kill leukemia and prostate cancer cells[3,4]. In addition to directly killing cancer cells, cannabinoids inhibit cancer through a surprising number of mechanisms. They stop cancer cell invasion, metastasis, growth, and proliferation, along with inhibiting angiogenesis to tumors[5,6,7]. These effects are produced through receptor-dependent and independent functions. For example, in some cases cannabinoid receptor activation triggers apoptosis, but in others there are different triggers. Cannabinoids even work at the genetic level, such as down-regulating the ID-1 gene in breast cancer[8].

Perhaps most importantly, there is abundant evidence that the human body is programmed to have cannabinoids kill cancer cells. Several studies show that endocannabinoids outright attack cancer cells[9,10]. Furthermore, Guzmán's pancreatic cancer study found that cancer cells had more cannabinoid receptors than regular cells. Combined with the fact that endocannabinoids kill cancer, it makes sense that the development of more cannabinoid receptors on cancer cells would be one of the body's innate defense mechanisms against the disease. Yet further, a study of people with liver cancer found that patients with higher levels of cannabinoid receptors had much better disease-free survival[11].

Looking at the experimental data, one might surmise that large doses of cannabis extracts would probably work against cancer in humans. After all, cannabinoids exert so many anti-cancer properties, and the human body is seemingly programmed to have cannabinoids kill cancer cells. Over the past six years, there has been a continuous stream of testimonials from individuals across the world, claiming that cannabis extract therapy eliminated their cancers. There are successful reports for virtually every type of cancer. Many people have provided full medical documentation online, including diagnosis and remission. The magnitude of cases cannot be ignored.

The caliber of entities making these reports is also enormously important. It's not just regular people - dispensaries, corporations, and doctors are saying it. Cannabis Science, a publically traded company, has documented two skin cancer remissions with topical application of extracts, and will soon be running clinical trials in Colombia. GW Pharmaceuticals is performing clinical trials of their cannabis medicine for brain cancer, and has patents on such use. Dispensaries like Harborside Health Center (the largest dispensary in the country) and River Rock have observed significant success with cancer treatment.

Dr. William Courtney and Dr. Jeffrey Hergenrather have extensively discussed their use of cannabis extract medicine to treat cancer in humans. Moreover, it was formally proven that cannabis extracts can kill leukemic cancer cells in at least one human patient in a November 2013 Case Reports in Oncology study, which even ruled out spontaneous remission and chemotherapy as causes of remission.

With millions of people suffering in hell-like conditions because cannabis extracts are illegal, I propose major reforms in cancer research priorities in 2015. The drug policy communities need to rally behind a re-prioritization of cannabis issues, putting cancer patients first. The urgency cannot be stressed, and I say it's time that the issue of the failure of medical and political infrastructure to confront cannabis science on the issue of treating cancers be more widely addressed.

Fear and Loathing in Amsterdam: The Smoke Abortion

Todd Brendan Fahey

A Self-Reflexive Fantasy/An Expressionist Argument...a trip down the Rabbit Hole: Is this the new face of Gonzo?; The New Paradigm Shuffle?...or Just a Hot New Way of Getting By?

"I have spent half my life trying to get away from journalism, but I am still mired in it -- a low trade and a habit worse than heroin, a strange seedy world full of misfits and drunkards and failures. A group photo of the top ten journalists in America on any given day would be a monument to human ugliness. It is not a trade that attracts a lot of 'slick' people; none of the Calvin Klein crowd or International jet set types. The sun will set in a blazing red sky to the east of Casablanca before a journalist appears on the cover of *People* magazine."

-- Hunter S. Thompson, *Generation of Swine: Tales of Shame and Degradation in the 80s*

[November 6, 1997 - email transmission from Fahey to R.U. Sirius]

Subject: 'Fear & Loathing Indeed'
-- but fr this place. P.U. Gotta boogie out of Louisiana SOON. Amsterdam was a dream. Just a dream. Where do I begin? OK: Fast-notation style: Finished the *Smoke* piece three days ago. It

blisters. Bought 4 pieces of original art, blew my *Smoke* $$$, but worth it. Got fantastic head from a Peruvian slut. wOw. Got meself into a rave/house music mag in the Netherlands -- *Basic Groove*: gonna be a meaty piece, & photos too! Am in this week's campus newspaper, back at the ranch. A good article; makes me look pretty paranoid, but I probably am. Got a good idea for a full length *Fear and Loathing in Amsterdam* novel. On which, more later. Smoked the kill hash; ate pure ecstasy (duzn't do much for me any more, but my friends tripped hard); a whammer LSD trip, rivaling anything I've ever been in touch w/; cubensis 3 times, 2 of them hardtrips. Peddled 35 copies of *Wisdom's Maw*: placed copies in: W.H.Smith (London-based); The Athenaeum Bookshop; The English Bookstore, & Conscious Dreams, a righteous head shop (took 15). Also got on TV. The broadcast was taped tonight (was on a plane home, but have a VHS); came on after Philip Glass (the composer). Yup. Pretty neat. *I TORE IT UP!* Was sure I was gonna lose my mind during days 2-6, but got it pretty well back together. I haven't used drugs like that in *years*. But FUN. The Red Light District is amazing. Too many sluts, too little money (actually, a 30 minute head job goes for 100 guilders, which is about $65...

"making it **hot** for them" - T. Southern-- tbf

There is almost no way to explain myself here, in the 2500 words allotted me by *Smoke*. I'm thankful like hell to have the chance to fill up a couple of their expensive pages, but it is a loaded prospect I face herewith. The enormity of the situation came to me on the plane ride over -- an uneventful 8 1/2 hrs through the sky -- a straight shot from Houston, the highlight of which was the Michael Keaton flop *Multiplicity*, a disturbing film about a slacker who can't advance his fortunes, no matter how many of himself he clones. I was thumbing through my hardback first-edition copy of *The Great Shark Hunt* (which I bought for a bargain $17 at some second-hand bookstore in Lafayette, and in which some sad-hearted fucker had written once: "To Nancy, the love of my life, 1979"...), culling what last-minute nuggets I could from my brutal Lord and Savior. It is an impossible act to follow, & I am too fundamentally honest to begin yammering about how the world really *needs* another Gonzo journalist around. The truth is, Hunter S. Thompson is a terrible genius, whose star is about to rise again. Soon there will appear a scholarly look at the mad Doktor's Life-Rant, and we will have him to kick around for another twenty or so years of internecine egghead warfare...which could make a decent segue into this Amsterdam piece.

A late-bloomer, I have been tracking Hunter Thompson since 1988--since on one sunny Santa Barbara afternoon and helplessly stoned on Humboldt, I was allowed to be taken in by his heavy con approach to the literary marketplace. Much can (and has & will yet) be said about Thompson's stylistic innovations, his "participatory journalism," yadda yadda. The thing about Thompson, for me, now, a 31-year old unknown novelist, is his understanding that a week on assignment in an exotic foreign locale is virtually always A Ticket To Ride. Open-ended gigs like this come around about as often as the comet Kahoutek. & there is simply no way of anticipating the kinds of connections, for good or ill, or both, that are to be made in the process of earning a heavy nonfiction Sex-&-Drugs legend.

There will be things told in this story that will scotch my reputation permanently in Puritan America. But the shitty truth of it is, Puritan America has never been particularly good to me. & there is also the fact that Hunter Thompson stopped writing serious Gonzo around 1979. The world has seen about nine major music movements since 1979. So maybe there *is* a need for another Gonzo journalist on the scene. Or maybe I'm just ego-stuffed and deranged...

My relationship with *Smoke* magazine may be important to clarify here. Assistant editor Rich Hoxey "got the call," as it were, when my felonious "acid novel" passed across his desk some time this past September, and is when Editor-in-Chief Aaron Sigmond signed off on the deal. *Smoke* was, at the time, devising a plan for a "Fear & Loathing" assignment, to be set in Amsterdam. They tell me I was third on the List: Chris Elliot had scheduling problems; Matthew Perry -- or one of those ubiquitous face-names...the *Friends* guy (& I'll check character assassination at the door)...some publicist for one among the *Friends* troupe decided that any story he wrote on or about Amsterdam would have to stink like Sex & Drugs, and that anything he wrote would be the death of his career; and so, alas, the duties fall to me. It is now *my* purpose in the Food Chain to bend the minds of the American reading populace. So let me begin.

There exists a world in which high-living men (and probably a couple of womyn) pay humans to get in trouble, or become stained by, and to confess publicly in venues such as *Smoke* (& before that, we had *Rolling Stone*, which is now the void that is Jann Wenner's existence...), there is a world in which writers are paid to experience things that could very well involve the bringing upon themselves heavy penal sentences. & once in a rare moon, there is a writer who, for whatever reason -- & they are all good and weighty -- has essentially "had it" with mainstream America: that vapid land of sitcoms and commercials & infomercials & talk shows & visionless, primarily materialistic subsistants & whatnot -- a writer who, again in the words of the great Hunter S., "has found out a way to live out there where the *real* winds blow."

So, I am the Acid Novelist. & I have been paid by *Smoke* to get The Story of Amsterdam. Let us proceed.

It is true, there exists a realm in which "happening minds" function more or less unfettered by meddling forces. That this place is called, geographically, Amsterdam, is also true. Many have found its gypsy soul/drunk of its wisdom. & maybe the real truth here is that America is not *worthy* of The Message. Maybe we've *blown* our shot. Maybe *that* is why Hunter has been so quiet for the past fifteen years. Indeed.

This is a strong line of inquiry, and it deserves to be plundered, & I am probably the fellow to do it; but there is also the issue of the Amsterdam piece, which may not be the most important thing on this writer's mind.

Hmm. Commerce is a heavy reality. There are *many* realities.

That is my message from Amsterdam. There exists a place where happening minds can be brought to beautiful (& probably terrifying) truths; where the body can be brought to pleasure in untold ways... There is a story here. Aaron Sigmond is getting the first whiff, because he laid down good money, proved himself a visionary fellow (or at least got really lucky).

There are portals into which happening minds can peer -- worlds into which, if one has balls enough, a man may find himself amongst splendidly amusing and generally very fine and even lucrative company. There is such a world. I call it Amsterdam.

Should this story be told to an American audience? The sadist in me sez, "Fuck 'em, they ain't worthy." But *Smoke* is paying. Commerce is a heavy reality. There are *many* realities. It is a solid paradox that is mine inhabitance.

Is America ready for me? Will it buy my acid novel? Can I make this gig pay? If so, I am the luckiest bastard alive. & Aaron Sigmond and his bosses above will prove themselves very good men -- like the last scions of the Medici giving funds to Michelangelo.

These are good shrooms.

I feel like Clint Eastwood in *Dirty Harry*: "In all this excitement, I lost count of how many rounds I've fired. Did I fire six shots, or did I fire five? Do you feel lucky, punk? Do you?" So, America: are you gonna stand by like suckers for another four years, while Bill-who-didn't-inhale sends beautiful bright minds to penitentiaries for seeking wisdom through chemicals? If so, you're no friend of mine.

The way I see it, it is time for many of us to make a deliberate, proactive choice: Revolt against the War on (some) Drugs...or move to Amsterdam; & if Amsterdam collapses as a place where happening minds can function fully and stay free...well, I will be in trouble. That would be a heavy day. I'll lay odds, though, now that I've been here and seen it for myself, that the Amsterdam intelligentsia would never let a thing like that happen.

So, I guess I'm an expatriate. Will any of my friends come over and play with me? & who will pay my bills? Bob Guccione? Hugh Hefner? Jann Wenner? *Or Aaron Sigmond*? My price is now $5,000 for a 5000-word installment of *Fear & Loathing in Amsterdam*, the novel. I think you're getting a bargain. Hunter won't get out of bed for less than twenty-five grand; and from what I've heard, from an agent we used to share together, Uncle Duke is now biding his daze in the fine company of Lady White and will calcify that way, more or less -- an exhilarating and disturbing fixture in the American psyche.

So, *wire that $$$, dear editors*, do it in Dutch guilders. Wire it to the Hotel Van Onna, 104 Bloemgracht. I stay in room 55. It is my lucky room. The proprietors know I'm trouble, but they take it with great humor. I have a twisted tale to share, bringing together a basement chemist named Heinrich, a smorgasboard of psychedelic shaman, a jazz player in exile since 1971 and an aging Gonzo journalist in need of spiritual redemption. It is a good story; I will put my soul into it. So, send the money, commissioning fellows. Allow me to finish this thing. Finance this pirate life of mine.

Do it now.

* * *

A "Fear & Loathing" place is a strange bird, journalistically. It lacks that which upon a true story depends: e.g., A Subject. And it is the lack of a subject that makes a "Fear & Loathing" piece as taxing as cleaning out the Augean stables. One lurches here and yon for an angle, casts about wildly for some goddamned Room with a View of Something Interesting...and when one is finally worn ragged -- because there is a God, and He is kind and has smiled upon his prodigal son on this day by granting him all three wishes in a single pop: an all-expense-paid trip to Amsterdam -- the philosopher's stone is delivered unto wild Gonzo man, such that the world might be brought to his arcane vision.

The first thing a journalist needs to know is his word-count -- in my case 2,500, which is, as you might guess, not a lot of leg-room for the long-boned storyteller. But it gives me leverage. Since the story I have been privileged with is a black diamond, Aaron Sigmond will either have to splay this bugger across the next half-dozen issues of *Smoke*, or I will have to publish another novel myself; & it is this win-win that brings to me the Mona Lisa Smile.

The other thing a Gonzo journalist needs to know is, how long his bills are being paid for. Aaron made it exquisitely clear that he was *not* paying for my Grand Tour, and I think I heard him toss off three days as a figure., which means I can stretch it to four -- kind of like going sixty-one in the old fifty-five m.p.h. scheme, and knowing you won't be made to suffer for it. But the injured Hell of it is that I had not a whole lot interesting *happen* to me in the first four days. The piece nearly got hijacked by a low-budget, garden-variety tale of a treacherous out-call girl and her pimply Iranian pimp/pusher boyfriend...but then the ecstasy started coming on, and revenge was no longer the best rush in town. And there was a night when I got involved in a savage fantasy featuring Courtney Cox and her body double, and I thought for a day or so that I might get some writing mileage out of that one.

It was around day seven when things began picking up the brilliant overtones unique to the Big Mystical Adventure. A quick mental survey told me it had been nearly a decade since my last significant sojourn in the weird world. Really good acid doesn't come around my neighborhood as often as it should.

The guarantors of *Smoke* will have many vital and well-founded questions about my trip to Amsterdam. It is unusual for a 2,500-word piece (which now looks like it will top 3k) to take eight days to accomplish; probably it is more unusual that such a trip breaks five figures in expenses.

Ah, let us burn yet another branch on the pyre of journalistic ethics.

Your money goes quickly in Amsterdam. Or, it has mine. I had been here 16 hours, and already I had spent $600: cab fare from the airport to the hotel @$45; a no-frills hotel in the museum district @$150/night; $200 for the hooker; $60 for two 3-gram baggies of hashish (not a lot, to the naked eye, but yr hash seems to stretch nice & far in Amsterdam...), another $60 for three baggies of freeze-dried psylocibin -- which is over-the-counter material (or at least until this article comes out) at head-shops Amsterdam-wide, and about $40 at a Transylvanian fish-house, whose manager put me immediately on edge. Persons from or around the Mediterranean, I've concluded, are the most provincial on earth -- a paranoiac, rigidly-suspicious strain, which probably has something to do w/ guarding the hallowed secrets of the Son of the Real Jehova. I read a good paperback on the subject once, got it in an airport -- but none of that matters now.

What I was after, after being forced by the autocratic young owner to clean my plate of the paella ("This is the *best* food," he kept repeating, "*Best* food in all Amsterdam. You don't like, we fix you something else. Come on, eat, this is *best food*..."), what I was after was the name of a good jazz club. I was in the mood to hear some young cat pluck off a stretch of guitar that would sound like Al DiMeola. It was a sophisticated mood, enhanced, no doubt, by the several pellets of 2cb I had eaten earlier in the evening.

"OK," I said, in a state of over-full exhaustion, and when he went to turn his well-oiled head to urge the Gen-X slacker waitress to earn her pay somehow, I managed to hide a couple boiled fish nuggets among or inside or underneath the mound of clam shells. "Is this good? Can I pay now?"

"Of course you can pay," he said, "You could have paid before. I just wanted you to eat *best food*.

I jogged across a cobbled walk to a smallish den called The Alto Club, on Liebenstraat, near the infamous Bulldog hash bar. I knew nothing of The Alto, except I had been warned that drink prices ran high for tourists. But at the door, I caught a vibe that was not real...*accommodating* to the traveling stranger: something about the way the clique of four pea-coated Dutchmen stared me down while I tried to ease my way through a clot of merrymakers. I actually thought about leaving, literally -- just turning around; but the band was called Gator's Groove, and mebbe I was feeling nostalgic for my old place in de bayou. Who fucking knows. I walked in, paid for a club soda, and got ready to make an end-run around what looked to be four bad-asses from the Holland countryside.

One of them made a sucking sound in the teeth. I remember being concerned that my laptop -- or the one I had borrowed from a friend -- would be damaged in the fall. And the bartender shook loose of the strap on my wrist. But when I went, I went.

The floors are all hardwood in Holland, and though hardwood has some give, it is not a lot, and the bones, it seems, are not conditioned to take an uncushioned freefall. I remember reading a *Reader's Digest* article on a stunt-diver who once fell 3,000 feet without a parachute, and lived. He broke most of the bones in his body, but had remarkably few internal injuries, and none to the brain. His tip was to fall on the pressure points on one's side: the shoulder, the elbow, the hip, the knee and the ankle. So that is what I did.

I took a long time to come to. When I finally did, it was midnight straight up, and I found myself seated at a high table, next to a large black man who wore green shades and looked to be from out of the cast of the Mod Squad. Quite suddenly, the foggy qualities of concussion had receded and I was aware of being in the presence of brother Lucius, of the 12th Panther Brigade, of Oakland. I was massaging the socket near my scapula, where the shoulder had been put back in two sure motions by someone whom many in the crowd called "the Healer," and who had since departed with little in the way of forwarding information.

Lucius laid down his glass of wormwood nectar. "Man, that's *nuthin'*," he said. His teeth were like cinder-blocks, very uniform and with all the grace of an old U-Haul building. The pores on his nose were huge and deep, like so many abandoned water wells. "You want me to tell you a thing like that shouldn't *happen* in this town? See, I know. *I* saw you comin'. Shit, this town's gon' be as mean as you want to *let* it be. They's cats here gib me the crawlies -- like findin' some old boy's head in yo' *bed* at night...no ketchup color on that picture, no *Godfather* hawse head, *shit.* But still, they's a spirit here *save yo' soul.* Save mine."

Lucius had seen better days, I knew. His eyes were clotted and rheumy, with real orange marmalade. They say absinthe is a harsh mistress, and he would make a good prohibition poster-child were it still a problem anywhere in the world.

He reached down into his lap and fumbled for something: I figured it was a cigar, but really I tried not to notice. Then, for the first time all night, I saw the glint of a tenor sax; the brass snake sat on the floor, its neck at rest against my companion's thigh, which was covered in a fine corduroy, of a rust complexion.

"Shit, most folks think of Amsterdam, and they see the steeple atop th' Temple of Gomorrah. But, they don't know. All I know is, a man can *think* straight over here. A man don't have to be scared all the time, 'bout gettin' his brains bashed in by some fool inbred thinks he's special 'cos he's Billy Joe's kissin' cousin, or some shit. I see what this place did for a lot o' sufferin' brothas. Saw Bud Powell go into a full bloom one May, right here, *right in this club*... If it wasn't this one, it was the one next door. And I don't have to tell some people what it's like to be 'round genius. Everything just kind'a *gives*," he said, pushing deliberately with his fingertips. "Everything becomes possible. It's like, before everything was walled off -- but you don't even know they's any walls there -- and once the genius hits, there's no more opposition. Everything's clear, and orderly. *Shit*," he glimmered. "It's *beautiful*."

TO BE CONTINUED...

Interview with Todd Fahey (TBF) in a long-ago Philly 'zine, Carbon 14 (C14)

C14: How long had *Wisdom's Maw* been completed before you decided to self-publish it?

TBF: *Wisdom's Maw* was conceived at the turn of 1988-89, in a period of increasingly heavy LSD usage for me. I was living in Santa Barbara, CA and had been accepted into the prestigious & ultra-expensive Professional Writing Program at USC for the Master's degree – this after having basically been run out of the Walter Cronkite School of Journalism at Arizona State University for writing "too much like Hunter Thompson." It is important to note that, at that time. (Spring 1988), I had never read Hunter Thompson. Not a word. I'm fairly sure I'd never even heard of him (I've spent many hours on this question and poured over all my old college folders. and there is no evidence at all of my ever encountering his work). I can pinpoint the exact moment that I recall discovering Thompson, and it was not until I bailed ASU in May of' '88 and moved back to Santa Barbara. (If you want to know the particulars, check out the article I wrote for a skin-mag back in '91, on my Web site.) This is an absolutely critical point if anyone is to fully appreciate my writing. I am not a "Hunter Thompson clone" (though one could do much worse). I share with Thompson the "black comedic" cast of mind; I am also an inveterate outsider and loner, with tendencies toward misanthropism.

CI4: What are the advantages you've found to being self-published?

TBF: Being self-published means basically to be a one-man (in my case) whole-service industry. I am the shipping clerk, the order taker, the PR mouth, the advertising specialist – the training for which has been all on-the-job. By the time I'm through with this whole *Wisdom's Maw* process, I really feel I could and should command a $100k yearly salary at a NY publishing house. 'Cause, if *Wisdom's Maw* takes off as a seller, it will be me & no other who got it there. The real advantage to me in this process, is to meet people like you. Truly, that's been the human bonus. I am on pretty cozy terms now with the ed/pubs of most of America's best counterculture magazines. No one knows better than I how fucking tough it is to get published, and to be able to just pick up the phone and say, 'Hey, buddy! I've got John Barlow hangin', wanna see the interview? Groovy. Let's do cola soon. Ciao," is kind of a mind-boggling thing. (I'm kidding about the cola – haven't touched it since 1987.) The horrendous *dis*advantage to self-publication is, obviously, the money. I'm in hock about $20k to Citibank right row and they ought to be conscious of remaining REALLY NICE TO ME, because it would be incredibly easy to declare Chapter 7 and call this whole thing a bad dream. But I want to keep Far Gone Books running. And so it's not in my best interest to go belly-up.

CI4: What influence have psychedelics had on your writing?

TBF: I have a deeply-embedded fear of being 'straight.' I'll be frank about it. I have been enamored of chemicals since my childhood and it is surely the bane of my existence. I lost my wife over it just this past year. I love her and respect her enough to have finally told her, 'I can't promise I will change & a promise is what you want.' So, we divorced after 5 1/2 years of a rewarding and tumultuous marriage. She did not know about my LSD intake during the writing of *Wisdom's Maw*. I hid it from her – an LSD addiction that sometimes went for 40 days in a row – and in hiding my usage, psychologically, I almost destroyed myself. I am digging myself out of the wreckage that is lies as we speak.

[For the record, it was my soon-to-be-ex-wife - just after I had told her about my LSD years - who laid out the book in Adobe PageMaker. She is a wonderful Mormon woman. I can't thank her enough. The getting out of this book is one of the miracles of modern medicine. Maybe someday I'll get a chance to write about it.]

My relationship with chemicals is an uncomfortable one. To be very honest I am either bored of the "sober life," or else it scares the shit out of me. I don't know which. From the age of seventeen, I don't think I've been straight more than a week at any given period. My survival is a testament to the strength of the human will. I had a hideous relationship with alcohol from 1982-1986 (from the age of 17 to a wizened 22, when I went through rehab.) I relapsed to the bottle in '93, after about the 120th rejection of *Wisdom's Maw*. I 'drank-to- die' until Thanksgiving of '95 – a fifth of Wild Turkey a day. I've shed many tears over the memory of those days. I was so desperate to get *Wisdom's Maw* published.

CI4: Is *Wisdom's Maw* your first book?

TBF: I've been writing a goddamned long time. I wrote my first book-length nonfiction 'novel' – a thing called *Hell Bottled Up: Chronicles of a Late Propaganda Minister* – in 1988, in my first semester at USC. Wrote it in a white-heat in six months, basically smashed on acid. *Hell Bottled Up!* is an autobiographical novel centering on my two violent years as a right-wing activist in Arizona, during the heyday of Governor Evan Mecham and a revival of the John Birch Society. You drop the name Todd Brendan Fahey around certain circles in Arizona today and you better watch your back. Oh, I've lived a really weird life.I became acquainted with conspiracy theory through the John Birch Society in 1984 and am credited with founding the first-ever college chapter of the JBS. I don't know whether to laugh or cry. But I was also a terrible drunk and was more than a little curious about psychedelics. Plus, I was a slut. (*Heeee.*)

That manuscript made the rounds of New York for three years, and at one point Faber & Faber fell in love with it, and Villard took a look at it. Thunder's Mouth Press wanted to see it specifically – but my then-agent couldn't close the deal. I finally shelved it in '92 , as I was becoming a better writer. I thought I should clean it up, and I didn't have the strength to look at it again. So, before I had even begun *Wisdom's Maw*, I had this other semi-notorious "novel" written and was frustrated about not selling it. So, by the time I got a hundred pages into this incredibly dark and deranged CIA/LSD novel, I was antsy to sell it pre-finish. I must have ridden my agent terribly. I was *so* certain it was going to be a blockbuster. I just could NOT understand why the NY majors weren't beating my door down. I still can't, *fools*! Now that *Wisdom's Maw* is getting great reviews in about every counterculture magazine that matter, I feel vindicated. It's not selling extremely well, but it's also not in very many book stores (like 10, maybe.) The chains won't touch it. I can't buy a distributor. It's a monster. I've sold about 3000 copies through my Web site and word of mouth. I unloaded 50 copies while I was in Amsterdam; visit any of the English/American book stores there and you'll find it. They loved it over there.

CI4: What other titles are planned for Far Gone Books?

TBF: *Fresh Fruit & Gravity*, a first-book of poems by Jim Tolan, will be out in about a month. It's a gorgeous thing, and at $9.95 (big commercial plug) is a steal, for a signed first. Jim is a friend of mine, a fellow Ph.D. at U. of Southwestern Louisiana, and a 1994 winner of the AWP Intro Award for poetry. He is working very much within the Whitman-bardic tradition – the larger "I-as-soul-of-America" thing – and I hope this book wins an award for best small press design, because it is stunning. I've been very lucky to have had two hungry graphics guys offer to design my first two books for pocket change. In May, I will release my demented short stories, titled *Dogshit Park & other atrocities*, which are the blackest things to come out since the heyday of Burroughs, Terry Southern and Hubert Selby, Jr., all of whom are my forefathers. After that, a collection of scholarly criticism on Hunter S. Thompson, which I think will surprise a lot of academics. Past that, I'll have to figure out my money situation. Any addled philanthropists out there reading this interview should mail checks & money orders to Far Gone Books' PayPal address: bill.rights@gmail.com

CI4: If you aren't trying to be the next Thompson, where did the idea of a "Fear and Loathing" piece originate?

TBF: As far as this "Fear and Loathing" piece, the story is pretty simple. I sent a review copy of *Wisdom's Maw* to *Smoke* (a NY cigar magazine aimed at Gen-X) and their assistant editor loved it. After a few fruitless phone calls back and forth with assignment ideas, they came up with the idea of "Fear and Loathing." I almost lost my lunch. Really. I walked around in a shit-eating daze for a week. So, I went to Amsterdam, started getting REALLY out of my head, like I hadn't in several years. (For the record, I stopped eating LSD in the summer of 1994 and, Bog willing, I will never pick up the habit again. Too many reminders. Too much psychic trauma. I'll probably do it again, 'cause I did it in Amsterdam – some incredibly pure & powerful stuff – but not as a "means of writing.")

But then it dawned on me: 'Crap, I can't pull a Hunter, Jr. I just can't.' I don't have much going for me these days – I'm probably unemployable in terms of a tenure-track teaching job, even though I will have my Ph.D. by May '97. God bless the school that gives me a gig. A "charitable institution," indeed. I have my writerly reputation and I can't afford to soil it. So, the article became a 'How-to-write a 'Fear and Loathing' piece," mixed with some insight on Thompson, who is my patron saint, and then a little segue into this fictional thing that will be *Fear and Loathing in Amsterdam: A Gonzo Novel*. Aaron Sigmond, editor-in-chief of *Smoke*, hated what I gave him, and he "killed the piece."

I admit, I went totally sideways on it; but I'm the loosest of cannons, and that's what I do best. My ex-wife loved that about me: "Never, *ever* a dull moment around the Toddmonster." It's going to be a great book and a lot of fun to write; but I like writing. I don't consider it, as does Hunter, "the most hateful kind of work." I'd rather be writing than doing just about anything – except maybe cruising the Red Light District of Amsterdam...so let me get back to work."

References

MDMA, PTSD And The Future Of Psychiatry

1.   **Bedi, G., et al., (2014)** A Window into the Intoxicated Mind? Speech as an Index of Psychoactive Drug Effects. Neuropsychopharmacology, 2014.
2.   **Brunner D, Hen R. (1997)** Insights into the neuro--- biology of impulsive behavior from serotonin receptor knockout mice. Ann NY Acad Sci 1997;836:81-105.
3.   **Cami J, Farré M, Mas M, et al. (2000)** Human pharmacology of 3,4-methylenedioxymethamphetamine ('ecstasy'): psychomotor performance and subjective effects. J Clin Psychopharmacol 2000;20:455-66.
4.   **Carhart---Harris RL, Erritzoe D, Williams LTJ, et al., (2014)** The Effects of Acutely Administered 3,4-Methylenedioxymethamphetamine on Spontaneous   Brain Function in Healthy Volunteers Measured with Arterial Spin Labeling and Blood Oxygen Level-Dependent Resting State Functional Connectivity, *Biological Psychiatry*, ISSN:0006-3223
5.   **Chabrol H. & Oehen, P. (2013)** MDMA assisted psychotherapy found to have a large effect for chronic post-traumatic stress disorder. J Psychopharmacol; s27(9):865-6.
6.   **Cozzi NV, Sievert MK, Shulgin AT, et al. (1999)** Inhibition of plasma membrane monoamine transporters by beta-ketoamphetamines. Eur J Pharmacol 1999;381:63---9.
7.   **Doblin R1, Greer G, Holland J, Jerome L, Mithoefer MC, Sessa B. (2014)** A reconsideration and response to Parrott AC (2013) "Human psychobiology of MDMA or 'Ecstasy': an overview of 25 years of empirical research". Hum Psychopharmacol. 2014 Mar;29(2):105-8. doi: 10.1002/hup.2389.
8.   **Fitzgerald JL, Reid JJ. (1990)** Effects of methylene-dioxymethamphetamine on the release of monoamines from rat brain slices. Eur J Pharmacol 1990;191:217-20.
9.   **Foa EB, Keane TM, Friedman MJ, Cohen JA. (2009)** Effective treatments for PTSD, practice guidelines from the International Society for Traumatic Stress Studies, 2nd edn. New York, NY: Guilford Press, 2009.
10. **Frye, C.G., et al. (2014)** MDMA decreases the effects of simulated social rejection. Pharmacol Biochem Behav, 2014. 117: p. 1-6.

11. **Graeff FG, Guimaraes FS, De Andrade TG, Deakin JF. (1996)** Role of
5-HT in stress, anxiety, and depression. Pharmacol Biochem Behav 1996; 54:129-41.

12. **Greer, G., & Tolbert, R. (1986)** Subjective reports of the effects of MDMA in a clinical setting. Journal of Psychoactive Drugs, 18 (4), 319-327

13. **Hysek, C.M., et al. (2013)** MDMA enhances emotional empathy and prosocial behavior. SocCogn Affect Neurosci, 2013.

14. **Kirkpatrick, M.G., et al., (2014a)** Effects of MDMA and Intranasal Oxytocin on Social and Emotional Processing. *Neuropsychopharmacology*, 2014.

15. **Kirkpatrick, M.G., et al., (2014b)** MDMA effects consistent across laboratories. Psychopharmacology (Berl), 2014.

16. **Lavelle A, Honner V, Docherty JR. (1999)** Investigation of the prejunctional alpha2-adrenoceptor mediated actions of MDMA in rat atrium and vas deferens. Br J Pharmacol 1999;128:975-80.

17. **Liechti ME, Vollenweider FX. (2000)** The serotonin uptake inhibitor citalopram reduces acute cardiovascular and vegetative effects of
3,4-methylenedioxymethamphetamine ('Ecstasy') in healthy volunteers. J Psychopharmacol 2000;14:269---74.

18. **Matthew Jakupcak, Matthew T. Tull, Michael J. McDermott, Debra Kaysen, Stephen Hunt, Tracy Simpson (2010)** PTSD symptom clusters in relationship to alcohol misuse among Iraq and Afghanistan war veterans seeking post-deployment VA health care Addictive Behaviors, Volume 35, Issue 9, September 2010, Pages 840–843

19. **Mithoefer et al (2013)** Durability of Improvement in PTSD Symptoms and absence of harmful effects or drug dependency after MDMA-assisted Psychotherapy: A Prospective Long-Term Follow-up Study. J Psychopharmacol, 2013. 27:28-39.

20. **Mithoefer MC, Wagner MT, Mithoefer AT, Jerome L ,Doblin R (2010)** The safety and efficacy of 3,4-- methylenedioxymethamphetamine-assisted psychotherapy in subjects with chronic, treatment-resistant posttraumatic stress disorder: The first randomized controlled pilot study. J. Psychopharmacol. 25(4): 439–452

21. **Nash JF, Roth BL, Brodkin JD, et al. (1994)** Effect of the R(-) and S(+) isomers of MDA and MDMA on

phosphatidylinositol turnover in cultured cells expressing 5-HT2A or 5-HT2C receptors. Neurosci Lett 1994;177:111-5.

22. **Oehen, P., et al. (2012)** A randomized, controlled pilot study of MDMA ({+/-}3,4-Methylenedioxymethamphetamine)-assisted psychotherapy for treatment of resistant, chronic Post---Traumatic Stress Disorder (PTSD). *J Psychopharmacol.*

23. **Roth, S.; Newman, E.; Pelcovitz, D.; Van Der Kolk, B.; Mandel, F. S. (1997).** "Complex PTSD in victims exposed to sexual and physical abuse: Results from the DSM---IV Field Trial for Posttraumatic Stress Disorder". *Journal of traumatic stress* **10** (4): 539–555.

24. **Schottenbauer MA, Glass CR, Arnkoff DB, Tendick V, Gray SH. (2008)** Nonresponse and dropout rates in outcome studies on PTSD: review and methodological considerations. Psychiatry. 2008 Summer;71(2):134-68

25. **Selvaraj, S. et al (2009)** Brain serotonin transporter binding in former users of MDMA ('ecstasy') BJP April 2009 194:355---359

26. **Sessa, B. (2012)** Could MDMA be useful in the treatment of PTSD? Progress in Neurology and Psychiatry. Volume 15, Issue 6, pages 4–7, November/December 2011

27. **Sessa, B. & Nutt, D.J. (2007)** 'MDMA, politics and medical research: Have we thrown the baby out with the bathwater?' J Psychopharmacol 21: 787-791.

28. **Thompson MR, Callaghan PD, Hunt GE, et al. (2007)** A role for oxytocin and 5HT(1A) receptors in the pro-social effects of 3,4, methylenedioxyamphetamine ('ecstasy'). Neuroscience 2007;146:509-14.

29. **Wardle, M.C. and H. de Wit, (2014)** MDMA alters emotional processing and facilitates positive social interaction. Psychopharmacology (Berl), 2014

Oriental Jones and the Medal of Freedom

1. In 1706, Sir William Jones Snr. (1675-1749), a Welsh navigator and mathematician, introduced π as shorthand for "the ratio of a circle's circumference to its diameter". He chose this Greek letter because it is the first letter in *periphery* and *perimeter*, Greek

words for circumference. Is use was popularized by Leonhard Euler, the great German mathematician, after Jones' death.

2. He was passionately fond of chess and when only seventeen composed the poem Caissa, an ode to the supposed muse or goddess of the game. Later in life he published *On the Indian Game of Chess* (1790), which includes a translation of a Sanskrit account of *caturanga*, the game's Indian precursor.

3. **The *Tariq-i-Nadiri*,** published as *l'Histoire de Nader Chah*, relates the true story of Nader Qoli Beg (1688-1747), a Turkish slave who became king of Persia.

4. **Stamets, P.**, *Psilocybin Mushrooms of the World, an Identification Guide*, Ten Speed Press, Berkeley. p.142.

5. *Ibid.*, p.143.

6. **Geoff Kibby**, editor of *Field Mycology*, in a personal communication.

7. Jones later published these opinions in *The Sanscrit Language* (1786).

8. **Hoffman, M. and Ruck, C. A. P.**, 2003, Freemasonry and the Survival of the Eucharistic Brotherhoods, Entheos Vol. III, Taos.

Cannabis and Alzheimer's Disease: Prevention and Treatment

1. **Hooper, C., Killick, R., Lovestone, S.,** "The GSK3 hypothesis of Alzheimer's disease" Journal of Neurochemistry, March 2008, 104(6): 1433-1439.

2. **Mackie, K., and Hille, B.,** "Cannabinoids inhibit N-type calcium channels in neuroblastoma-glioma cells." Proceedings of the National Academy of Sciences for the United States of America, May 1 1992, 89(9): 3825-3829.

3. **Hampson, AJ, et al.,** "Cannabidiol and (-) Delta9-tetrahydrocannabinol are neuroprotective antioxidants." Proceedings of the National Academy of Sciences of the United States of America, (July, 1998): 8268-73.

4. **Luvone, T, Esposito, G, Esposito, R, et al,** "Neuroprotective effect of cannabidiol, a non-psychoactive component from Cannabis sativa, on beta-amyloid-induced toxicity in PC12 cells." Journal of Neurochemistry, April 2004, 89(1): 134-141.

5. **Eubanks, LM, Rogers, CJ, Beuscher, AE** 4th, et al, "A molecular link between the active component of marijuana and Alzheimer's disease pathology." Molecular Pharmaceutics, November-December 2006, 3(6): 773-777.

6. **Martin-Moreno, AM, Brera, B, Spuch, C, et al,** "Prolonged oral cannabinoid administration prevents neuroinflammation, lowers B-amyloid levels and improves cognitive performance in TG APP 2576 mice." Journal of Neuroinflammation, January 2012, www.ncbi.nim.nih.gov/pmc/articles/PMC3292807. (accessed July, 24, 2014.)

7. **Chen, R, Zhange, J, Fan, N, et al,** "Delta-9-THC-Caused Synaptic and Memory Impairments Are Mediated through COX-2 Signaling," Cell, November 2013, www.cell.com/abstract/S0092-8674(13)01360-3.

8. **Casarejos, MJ, Perucho,J, Gomez, A, et al,** "Natural cannabinoids improve dopamine neurotransmission and tau and amyloid pathology in a mouse model of taupathy." Journal of Alzheimer's Disease, 2013, 35(3): 525-539.

9. **Ramirez, BG, Blazquez, C, Gomez del Pulgar, T, et al,** "Prevention of Alzheimer's disease pathology by cannabinoids: neuroprotection mediated by blockade of microglial activation," Journal of Neuroscience, February 2005, 24(8): 1904-1913.

10. **Van der Stelt, M, Mazzola, C, Esposito, G, et al,** "Endocannabinoids and beta-amyloid-induced neurotoxicity in vivo: effect of pharmacological elevation of endocannabinoid levels, Cellular and Molecular Life Sciences" June 2006, 63(12): 1410-1424.

11. **Cao, C, Li, Y, Bai, G, et al,** "The Potential Therapeutic Effects Of THC on Alzheimer's Disease." Journal of Alzheimer's Disease, July 2014,.

12. **Nadler, V, Mechoulam, R, Sokolovsky, M,** "The non-psychotropic cannabinoid (+)-(3S,4S)-7-hydroxy-delta-6tetrahydrocannabinol 1,1-dimethylheptyl (HU-211) attenuates N-methyl-D-aspartate receptor-mediated neurotoxicity in primary cultures of rat forebrain." Neuroscience Letters, November 1993, 162(1-2):43-45.

13. **Grotenhermen, F,** "Cannabinoids," Current Drug Targets. CNS and Neurological Disorders, October 2005, 4(5): 507-530.

14. **Cheng,X, Shen, Y, Li, R,** "Targeting TNF: a therapeutic strategy for Alzheimer's Disease," Drug Discovery Today, July 2014.

15. **Saco, T, Parthasarathy, PT, Cho, Y, et al,** "Inflammasome: a new trigger of Alzheimer's disease," Frontiers in Aging Neuroscience, May 2014.

16. **Nargakatti, P, Pandey, R, Reider, SA, et al,** " Cannabinoids as novel anti-inflammatory drugs," Future Medicinal Chemistry, October 2009, 1(7): 1333-1349. 17. **Rajavashisth, TB, Shaheen, M, Norris, KC, et al,** "Decreased prevalence of diabetes in marijuana users: cross-sectional data from the National Health and Nutrition Examination Survey (NHANES) III." BMJ Open, February 2012,
18. **Brookmeyer, R, Gray, S, Kawas, C,** "Projections of Alzheimer's Disease in the United States and public health impact of delaying disease onset," American Journal of Public Health, September 1998, 88(9): 1337-1342.
19. **Volicer, L, Stelly, M, Morris, J, et al.,** " Effects of dronabinol on anorexia and disturbed behavior in patients with Alzheimer's disease." September 1997, 12(9): 913-919.
20. **Walther, S, Mahlberg,R, Eichmann, U, Kunz, D,** "Delta-9-tetrahydrocannabinol for nighttime agitation in severe dementia." Psychopharmacology, May 2006, 185(4): 524-528.

The Oncological Evidence For The Effective Use Of Cannabis Extracts In Treatment Of Various Cancers

1. Delta9-tetrahydrocannabinol induces apoptosis in C6 glioma cells.
http://www.ncbi.nlm.nih.gov/pubmed/9771884
2. Cannabinoids Induce Apoptosis of Pancreatic Tumor Cells via Endoplasmic Reticulum Stress–Related Genes.
http://cancerres.aacrjournals.org/content/66/13/6748.full
3. Delta9-tetrahydrocannabinol-induced apoptosis in Jurkat leukemia T cells is regulated by translocation of Bad to mitochondria.
http://www.ncbi.nlm.nih.gov/pubmed/16908594
4. Delta9-tetrahydrocannabinol induces apoptosis in human prostate PC-3 cells via a receptor-independent mechanism.
http://www.ncbi.nlm.nih.gov/pubmed/10570948
5. Cannabidiol inhibits lung cancer cell invasion and metastasis via intercellular adhesion molecule-1.
http://www.ncbi.nlm.nih.gov/pubmed/22198381
6. Cannabidiol, a Non-Psychoactive Cannabinoid Compound, Inhibits Proliferation and Invasion in U87-MG and T98G Glioma Cells through a Multitarget Effect.

http://www.plosone.org/article/info%3Adoi%2F10.1371%2Fjo
urnal.pone.0076918
7. Cannabidiol inhibits angiogenesis by multiple mechanisms.
Http://www.ncbi.nlm.nih.gov/pubmed/22624859
8. Cannabidiol as a novel inhibitor of Id-1 gene expression in
aggressive breast cancer cells.
http://mct.aacrjournals.org/content/6/11/2921.long
9. Anti-proliferative and apoptotic effects of anandamide in
human prostatic cancer cell lines: implication of epidermal
growth factor receptor down-regulation and ceramide
production.
http://www.ncbi.nlm.nih.gov/pubmed/12746841
10. Anandamide Induces Apoptosis in Human Cells via Vanilloid
Receptors.
http://www.jbc.org/content/275/41/31938.full
11. Overexpression of cannabinoid receptors CB1 and CB2
correlates with improved prognosis of patients with
hepatocellular carcinoma.
http://www.ncbi.nlm.nih.gov/pubmed/17074588

Author Bios

Neal M. Goldsmith, Ph.D. is a psychotherapist and author specializing in psychospiritual development. He is also a public speaker, and curates and hosts innovative workshops, salons, and conferences on psychedelic therapy, innovation and change, and the post-modern future of society. Dr. Goldsmith's book, *Psychedelic Healing: The Promise of Entheogens for Psychotherapy and Spiritual Development*, describes the influence of psychedelics on the development of his personality theory and clinical practice. Trained in humanistic, transpersonal, and eastern traditions, Dr. Goldsmith maintains a (non-psychedelic) psychotherapy practice in New York City and may be reached via his Web site:
http://www.nealgoldsmith.com/psychedelics.

Tristan Gulliford is a writer, musician, and game designer who graduated in 2010 with a double major in Religious Studies and Literature, writing an Honors Thesis titled *Entheogens and Mysticism: Perspectives on Altered Consciousness from Religious and Scientific Worldviews.* Tristan helped create Evolver magazine and was an Associate Editor before it became Reality Sandwich. He is webmaster for Psychedelic American.

Dr. Ben Sessa, MD, is a medical doctor working in the field of addictions. For the last ten years he has been researching the potential role for psychedelic drug-assisted psychotherapy in the treatment of a range of mental disorders. He is particularly interested in developing MDMA as a treatment for PTSD. Ben is one of the co-founders and chairs of the UK conference on psychedelic consciousness research, *Breaking Convention* and author of *The Psychedelic Renaissance*. He also likes to brew cider and play the trumpet in Bristol's favourite Hip Hop Ska band, *Fat Sandwich.*

Todd Brendan Fahey is author of *Wisdom's Maw* (Far Gone Books, 1996) and *Dogshit Park & other atrocities* (Far Gone Books, 2014); his novel-in-progress, A String of Saturdays: The New Southern Romance, is slated for spring 2017 release. His published interviews include Ken Kesey, Timothy Leary, John Perry Barlow, John Shirley, R. U. Sirius and Douglas Rushkoff.

Angel Majao is a neuroscience student and independent researcher in his third year at Brooklyn College. He is the chief assistant to the science editor at Psychedelic American.

Randy Sloane (RS Aaron) is a peace-loving truth-seeking individual who received his degree in communications from the University of Maryland, College Park in 2012 and is currently aspiring to find more challenging work in the psychedelic research community, whilst simultaneously earning his feathers as the copy editor at *Psychedelic American.*

Benjamin Stolz is a student and researcher from central New Jersey currently studying anthropology and religion at Rutgers, New Brunswick. He is the founder of *Psychedelic American.*

John Hoopes, Ph.D, is an archaeologist who lives in Lawrence, Kansas who is an expert on a number of diverse Mesoamerican civilizations' cultures and their influence on New Age syncretic religions. He is currently a professor of archaeology at the University of Kansas.

Kevin Whitesides is currently a Ph.D student at the University of California, Santa Barbara. A religious scholar, Whitesides specializes in understanding the 2012 meme and its impact on both popular culture of the 20th century and the counterculture of the 1990s. He lives with his wife in Goleta, CA and is always looking for rare and obscure books for his library of primary and secondary works on the subject(s) of the occult.

James W. Jesso is a Canadian author and public educator who is dedicated to helping to engender regenerative, accessible, and public discourse on socially outcast and taboo subjects. He has authored numerous articles, essays, and videos, done radio, podcast, and print interviews, and toured across the world giving lectures and facilitating public discussions on a variety of subjects. Jesso has given over 40 public presentations since the start of 2013. Most recently, he has taught in Peru on plant-based psychotherapy while touring the jungle, studying and exploring the use of Ayahuasca as a spiritual tool. He has published two books, *Decomposing The Shadow: Lessons From The Psilocybin Mushroom* (2013) and *Soundscapes & Psychedelics* (2014). His third book, *The True Light Of Darkness,* is a deep and vulnerable look into the entheogenic experiences that brought him out of depression and into his creative passions. It is scheduled for release in the summer of 2015. Check out more of Jesso's work through http://www.jameswjesso.com/.

Justin Kander is the former webmaster of PhoenixTears.ca and an advocate for the use of cannabis extract medicine to treat cancer. He previously interned with Students for Sensible Drug Policy, a drug policy reform organization. In October 2013, Kander completed the *Comprehensive Report on the Cannabis Extract Movement* (CannabisExtractReport.com), a 100-page report that examines and analyzes the mounting evidence in support of cannabis extracts as medicine. He informally shared his findings at the International Drug Policy Reform Conference in Denver the same year. The report is now in its 7th edition and is 227 pages. Justin also presented about cannabis and cancer in November 2014 at the *Inaugural Australian Medicinal Cannabis Symposium* and released *Enhancing Your Endocannabinoid System* in February 2015, which includes natural ways for anyone to improve the health of their endocannabinoid system and increase effectiveness of cannabis medicine. The latter report can be found at: http://EnhancingYourECS.com.

Clint Werner is a cannabis researcher from the Seattle area. He is the author of *Marijuana: Gateway to Health* but works full-time as a show breeder specializing in Dachshunds. He has recently been featured on C-SPAN discussing his work on cannabis' psychological and physiological benefits.

Mike Crowley was born in Wales. A chance meeting with a Tibetan lama of the Kagyud order led him taking the vows of a Buddhist layman in 1970. After seventeen further years of study and practice, he was ordained as a lama. In addition to his studies of the Classical Tibetan language, he has also studied Sanskrit and Chinese. He taught The Epistemology of the Abhidharma at the Jagellonian University, Krakow, and lectured on Tibetan history at the Museum of Asia and the Pacific, Warsaw. He has given talks at the Polish National Buddhist Center and at the Center for Integral Studies (San Francisco) and presented at various conferences. His writings include an online book, *Secret Drugs of Buddhism*, and a number of essays, including *The Abduction of Tara* and *When Gods Drank Urine*. A paper entitled *Umbrellas, Wheels and Bumps on the Head* is the cover article of the June 2015 issue of Time & Mind.

www.ingramcontent.com/pod-product-compliance
Lightning Source LLC
Chambersburg PA
CBHW070832180526
45168CB00002B/811